Autorenkollektiv

Plasma Chemistry

The Mechanism and Kinetics of Plasma Polymerization

A. T. Bell

Plasma Chemistry of Fluorocarbons as Related to Plasma Etching and Plasma Polymerization

E. Kay, J. Coburn and A. Dilks

Elementary Processes at Solid Surfaces Immersed in Low Pressure Plasmas

H. F. Winters

AKADEMIE-VERLAG · BERLIN

Autorenkollektiv

PLASMA CHEMISTRY

(Editors: S. Vepřek, M. Venugopalan)

AKADEMIE-VERLAG · BERLIN
1981

Die Originalausgabe erscheint im Springer-Verlag
Berlin · Heidelberg · New York
in der Schriftenreihe „Topics in Current Chemistry",
Volume 94

Vertrieb ausschließlich für alle Staaten mit Ausnahme der sozialistischen Länder:
Springer-Verlag Berlin · Heidelberg New York
Vertrieb für die sozialistischen Länder:
Akademie-Verlag Berlin

Erschienen im Akademie-Verlag, 1080 Berlin, Leipziger Straße 3—4

Lizenznummer: 202 · 100/517/81
Gesamtherstellung: VEB Druckerei „Thomas Müntzer", 5820 Bad Langensalza
Umschlaggestaltung: Karl Salzbrunn
Bestellnummer: 762 940 4 (6619) · LSV 1215
Printed in GDR
DDR 72,— M

Springer-Verlag, Postfach 105280 · D-6900 Heidelberg 1, Telephone (06221) 487-1 · Telex 04-61723
Heidelberger Platz 3 · D-1000 Berlin 33, Telephone (030)822001 · Telex 01-83319
Springer-Verlag, New York Inc., 175. Fifth Avenue · New York, NY 10010, Telephone 477-8200

Contents

Plasma Chemistry of Fluorocarbons as Related to Plasma Etching and Plasma Polymerization

Eric Kay, John Coburn and Alan Dilks

IBM Research Laboratory, San Jose, California 95193

This chapter deals with the physical and chemical processes which take place in fluorocarbon plasmas. Emphasis is given to plasma diagnostic techniques which are well suited to investigate the neutral and ionized gas phase species as well as plasma-surface interactions. Special attention is given to a description of mechanisms which control plasma etching and plasma polymerization. Structural characterization techniques which work well with a variety of fluorocarbon thin films prepared in a plasma are given and key physical properties of such films are reviewed.

Table of Contents

1 General Considerations of Fluorocarbon Plasmas

The renewed impetus to studying fluorocarbon discharges is clearly driven by the needs of the rapidly emerging etching and polymerization technologies which require on the one hand the ability to selectively remove material from a surface with ever increasing directional resolution as needed in fine line lithographic applications and on the other hand to deposit polymeric material uniformly over large surface areas. Numerous investigations directed towards obtaining a better understanding of the fundamental physical and chemical processes which take place in the complicated glow discharge environments used for these applications have been initiated in several laboratories. The problem is a dual one in that it is important to understand not only the dominant gas phase collisional processes, but also the consequences of all the major gas phase species colliding with a surface in the discharge. Clearly, the etching and polymerizing behavior is determined ultimately by particle-surface interactions.

Recent in situ plasma diagnostic studies involving halocarbon discharges have helped considerably in delineating some of the physical and chemical processes. It is clear that most of the important plasma gas phase excitation, ionization and fragmentation processes in the 10^{-1}–10^{-4} Torr pressure regime are primarily controlled by collisions of the gaseous species with electrons, whereas the collisional partners which lead to subsequent loss processes of active gaseous species are manifold. It is self evident therefore that control over electron energy and density in various regions of the plasma will afford considerable control over the production of highly active species which are thought in part to be responsible for some of the unique reactions encountered in plasmas. There are many well documented ways, especially in the related thin film sputtering literature, in which electron energy and density in various discharge modes can be manipulated. Such control of electron density and energy is highly desirable to obtain reproducibility in various discharge configurations, but as yet does not in itself allow us to accurately predict which species will ultimately survive in a multicollisional molecular fluorocarbon discharge. Further mention of this subject will appear later in this chapter.

Emphasis will also be placed on approaches which lead to the removal of reactive species from the gas phase as well as the special role of energetic positive ions in plasma-surface interactions. Controlled scavenging of critical species from the gas phase and/or at specific surfaces and the degree of positive ion bombardment at a given surface can in fact result in simultaneous polymerization at one surface and etching at another within the same apparatus.

Attention will also be drawn to the significance of the residence time of active species in the plasma region of interest. It will become evident that the flow rate and pressure as well as the consumption of active species per unit area of active surface can alter the resultant chemistry dramatically. Clearly, one should expect both the etching and polymerization processes to be different depending on where the polymerizing or etching surface of interest is located with respect to the plasma region. One would expect and find that, for example, the polymerization process will approach conventional polymerization mechanisms and resultant structures the further removed the polymerizing surface of interest is from the active plasma region.

It is also important to recognize that unique differences should be anticipated in the plasma chemistry of fluorocarbons from that of hydrocarbons. This can be attributed in part to the strong C—F bond and relatively weak F—F bond. Thus, while elimination of H_2 and HF is a favorable process in the case of hydrocarbons and fluorohydrocarbons, for fluorocarbons C—C bond cleavage is more likely. For example:

$$CH_2{=}CH_2 \xrightarrow{e^-} CH{\equiv}CH + H_2 \qquad \Delta H = 41.7 \text{ kcal/mole}$$
$$CH_2{=}CHF \xrightarrow{e^-} CH{\equiv}CH + HF \qquad \Delta H = 18.2 \text{ kcal/mole}$$
$$CF_2{=}CF_2 \xrightarrow{e^-} CF{\equiv}CF + F_2 \qquad \Delta H = 162.4 \text{ kcal/mole}$$

$$CF_2{=}CF_2 \xrightarrow{e^-} 2\,{:}CF_2 \qquad \Delta H = 70.4 \text{ kcal/mole}$$
$$CH_2{=}CH_2 \xrightarrow{e^-} 2\,{:}CH_2 \qquad \Delta H = 171.7 \text{ kcal/mole}$$

The much greater stability of difluorocarbene than methylene allows it to undergo many collisions in the gas phase without reaction, and has therefore been found to be a prominent specie in plasmas excited in fluorocarbons. In contrast to this, plasmas excited in hydrocarbons are often dominated by a high partial pressure of hydrogen.

Whereas many discharge configurations are being explored for various etching and polymerization applications, the rf capacitively coupled, planar diode systems appear to be emerging as the most reliable and flexible to meet these various technological needs. Furthermore, the broad experience gained by many from the much more mature sputtering technology using this configuration provides considerable insight into the significance and control of the relevant discharge parameters. Consequently, much of the discussion in this chapter will apply directly to processes prevailing in such diode configurations, but obviously can be qualitatively extrapolated to other configurations which will also be mentioned.

Another key point to recognize is the fact that in rf driven diode configurations, as commonly used, all surfaces including the grounded surfaces are at a negative potential with respect to the plasma potential and are therefore subject to positive ion bombardment. The negative bias voltage at the rf driven excitation electrode is determined primarily by the amplitude of the capacitively applied rf voltage and the relative area of this electrode to that of all the grounded surfaces. On the other hand, only the relative surface area of all the grounded surfaces and electrode position dictate the energy with which positive ions arrive at such surfaces. The energy of positive ions incident on a given surface which has been isolated electrically from ground (possibly substrates) can, however, be independently controlled by applying a predetermined dc negative potential to such surfaces from an external power source. Certainly at the negatively biased excitation electrode negative charge carrier bombardment is insignificant and at all other surfaces only those negative charge carriers (electrons and negative ions) which have sufficient energy to overcome the sheath potential can ultimately arrive at that surface. Sheath potentials at grounded surfaces in commonly used of diode configurations can vary from a few tens to several hundred electron volts depending on the particular geometry. Consequently, the role of negative charge carriers in the surface plasma etching and surface polymerization process is considered secondary especially in the higher pressure regimes and will not be emphasized in this chapter. The role of negative ion gas phase collisions may, however, be quite significant but to our knowledge definitive evidence as to their role has not been established so far in this context.

We shall see, the particular surface chemistry which ultimately dominates at a particular surface in a given experiment will be greatly influenced by the flux and energy of positive charge carriers that arrive at that surface simultaneously with ground state and excited state neutral species as well as the chemical nature of the surface.

A brief description of the key plasma diagnostic techniques, which have been especially useful in delineating the gas phase processes in fluorocarbon plasmas, will be given followed by an extensive discussion of plasma etching and polymerization mechanisms.

2 Experimental Methods Used in the Study of Etching and Polymerizing Plasmas

2.1 Plasma Characterization

It is far beyond the scope of this chapter to consider the general field of plasma diagnostics[1-3]. Instead we will discuss briefly those specific diagnostic methods which have been used to study etching or polymerizing plasmas and we will speculate on the value of the information which might be derived from other diagnostic procedures. One of the major problems encountered by workers in the field of plasma chemistry is the large number of parameters which influence the process. This parameter problem is illustrated in Fig. 2.1 in which the interactions between the numerous readily accessible parameters and the more basic plasma parameters and the plasma-surface interactions are emphasized. It is clear that one cannot expect experimental results to be reproducible from one plasma system to another because of this extensive as yet unstandardized parameter space.

2.1.1 Electron Density N_e and Energy Distribution f(E)

Prior to initiating a plasma characterization experiment, it is important to delineate clearly what information is needed. For example, in most plasma etching and plasma polymerization situations, the important interactions are the consequences of the plasma-surface interaction. The key insight that is usually needed in these studies is the identity, abundance and energy of all species incident on the surface of interest. This information is not easily obtainable from just a knowledge of the electron density N_e and energy distribution f(E), and so far relatively little effort has been devoted to determining these quantities in etching and polymerizing

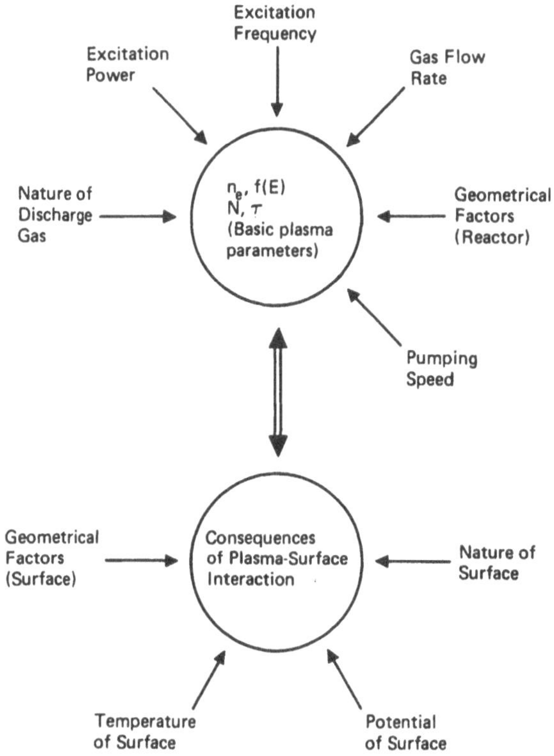

Fig. 2.1. Representation of the parameter problem in plasma-surface interaction. n_e-electron density, f(E)-electron energy distribution, N-gas density, τ-residence time for gas molecules in plasma volume

discharges[4)]. These quantities have been used to characterize plasmas of atomic gases but the complexity of molecular gas discharges, particularly when the discharge gas is triatomic or larger, would appear to preclude quantitative information about species concentrations being derived from N_e and f(E) since much of the cross section data is not yet available.

As more basic information about ionization cross sections[5-7)], dissociation cross-sections[8-10)] and cross-sections for the numerous ion-molecule[11-14)] and radical-molecule[15)] collisions become available, it may become possible to extract qualitative information concerning the density of all the various ionic and neutral species in molecular gas discharges. Steps in this direction are being taken with molecular gas laser discharges but the utility of this approach has not, to our knowledge, been demonstrated in etching and polymerizing discharges. It should be pointed out, however, that a knowledge of N_e and f(E) is expected to be very useful in the scaling up of etching and polymerizing processes in different discharge configurations. The complexity of the parameter space as emphasized in Fig. 2.1 causes serious problems in scaling up processes which have been developed in small research systems and a direct measurement of N_e and f(E) will help establish the operating parameters of a new system.

There are several methods by which N_e and in some instances f(E) can be determined experimentally[1-3)]. However, none of these methods can be thought of as easily applicable to etching and particularly polymerizing plasmas. The potentially most useful method would appear to be electrostatic probes[17–21] but even this method is fraught with experimental and interpretational difficulties[4,22–25] particularly under polymerizing conditions. Therefore, one can offer two reasons for the scarcity of N_e and f(E) determinations in etching and polymerizing plasmas — the fact that a knowledge of these quantities does not provide an answer to the key questions being asked and the fact that a reliable measurement of N_e and f(E) in these fluorocarbon plasmas is non-trivial. In this context, some recent promising results[16] have been obtained using optical emission spectroscopy of CF_4 discharges to which a small known amount of argon ($\sim 2\%$) has been added. The emission intensity of the Ar lines provides a measure of the $N_e \cdot f(E)$ product for electrons with energies above the threshold for forming the excited state of argon from which the optical signal originates. It is anticipated that optical emission spectroscopy, used with an "actinometer" such as argon, will be able to help overcome the parameter problem in plasma processing.

2.1.2 Plasma Potential

The importance of energetic particle bombardment on gas-surface chemistry is well-established[26] and the mechanisms involved will be discussed by H. F. Winters in a following chapter. Consequently, it is important to know as much as possible about the energy of ions and electrons incident on surfaces of interest in etching and polymerizing plasmas. This necessitates a knowledge of the potential of the surface V_s as indicated in Fig. 2.1 but also requires a knowledge of the plasma potential V_p since the energy of positive ions incident on surfaces is determined by V_p–V_s modified by collisional processes in the plasma-surface boundary sheath. The plasma potential can be determined from the I–V characteristics of electrostatic probes[17,21,27] albeit with some uncertainty. The floating potential cannot be used to estimate the plasma potential in rf discharges where the plasma potential is often determined by a capacitive division[28,29] of the rf voltage applied to an internal electrode. In these systems the plasma potential depends upon the relative area of the excitation electrode and of other grounded surfaces in contact with the discharge[28,29] and can be as large as a few hundred volts positive with respect to ground when this area ratio approaches unity. An example is shown in Fig. 2.2. Floating potentials in these systems are usually near ground potential[29]. The plasma potential can be determined by energy analysis of positive ions extracted directly from the discharge[29]. This approach can provide accurate measurements of the plasma potential if collisional processes in the plasma-extraction orifice boundary are considered, but the instrumentation needed is sufficiently complex to preclude widespread use of this method. The plasma potential can be seriously influenced by the presence of a third electrode, positively biased, as is shown in Fig. 2.3 in rf capacitively coupled discharges[29] or by a dc grounded excitation electrode. It is suspected that the lack of information concerning plasma potentials has led to some confusion concerning the extent of energetic positive ion bombardment in commercially available plasma reactors[30] as pointed out by Vossen.

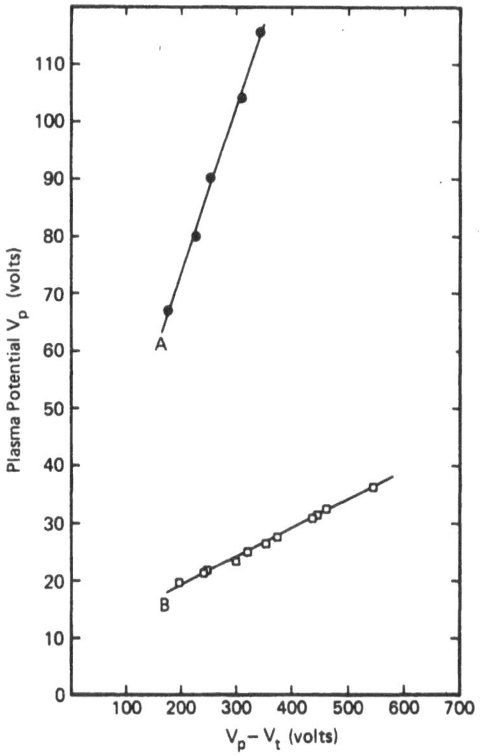

Fig. 2.2. Plasma potential V_p vs the dc voltage across the target-plasma sheath ($V_p - V_t$) for confined and unconfined discharges. In curve A (confined) the target area is 28.9 percent of the area of non-target surfaces in contact with the discharge whereas in curve B (unconfined) the target area is only 9.2 percent of non-target areas. The discharge was established by capacitively applying 13.56 MHz rf power to the target electrode in the presence of 50 millitorr of argon. The inter-electrode spacing (target-ground plane) was 1.88 cm. (See Ref. [29])

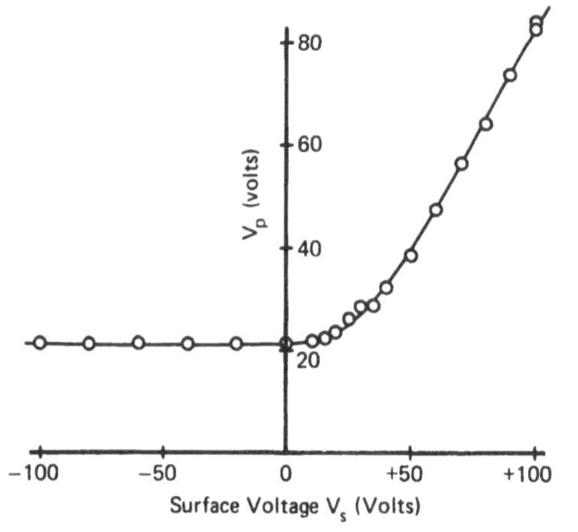

Fig. 2.3. Plasma potential V_p vs the dc voltage applied to a small surface ($\sim 1.3\ cm^2$) in the discharge. Argon pressure = 75 millitorr, 13.56 MHz rf power = 100 watts, target area = 100 cm^2, inter-electrode spacing = 5 cm. (See Ref. [29])

2.1.3 Gas Flow Parameters

The specification of gas flow parameters should always include a measurement of the total pressure p and the gas flow rate Q. This is easily done with simple instrumentation yet much data appears in the literature with only one of these parameters specified. The residence time of a molecule in the plasma zone τ can be expected to be an important parameter in plasma chemistry involving molecular gas discharges. It is therefore necessary to specify, as best as one is able, the volume of the plasma V. If this is done, τ can be determined directly from the relation $\tau = pV/Q$. Also, it should be emphasized that when p is varied as a parameter, either Q or τ should be kept constant and specified. Often it is not clear which of these two methods of varying p is to be preferred. An example of this problem is shown in Fig. 2.4 in which the pressure dependence of the CF_4 plasma etch rate of silicon is shown for the two methods of CF_4 pressure variation[31)]. The enormous difference observed in this example is attributed to the fact that the etching process consumes a substantial fraction of the injected fluorine and at constant flow rate this fluorine consumption has a smaller effect on the plasma chemistry than when the flow rate is varied, particularly for low flow rates where the utilization factor[32)] is large.

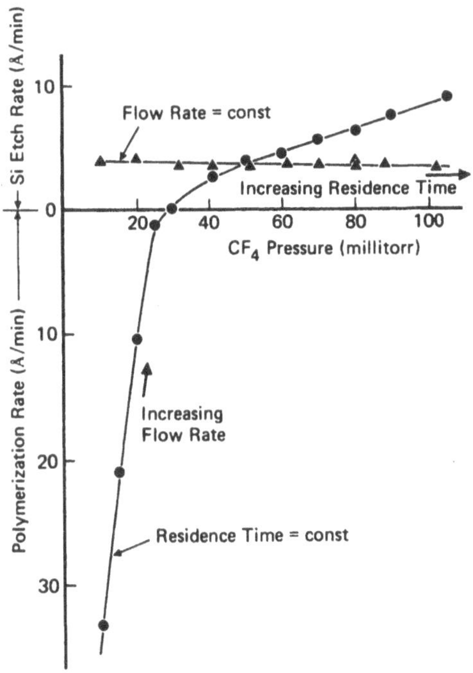

Fig. 2.4. Etching and polymerization dependence as a function of pressure in a CF_4 plasma. Pressure varied by the different techniques involving flow rate and residence time

2.1.4 Species Identification and Measurement

As was mentioned earlier, one of the most important pieces of information that is needed to develop an understanding of etching and polymerizing processes is the nature and abundance of all species incident on the surface of interest. Quite naturally, therefore, one finds that there are many studies directed towards identifying and quantifying various species in etching and polymerizing discharges.

The two approaches which are used almost exclusively to accomplish this task are mass spectrometry and optical spectroscopy.

Mass spectrometric techniques have been used extensively to analyze the species present in glow discharges. The motivation is usually either to obtain mechanistic insight related to plasma processes or to determine cross-sections for specific reactions (usually ion-molecule collisions) in well-defined plasma conditions, usually in a spatial or temporal afterglow. Two basic approaches can be used: 1) Direct sampling of the ionic species from the discharge, and 2) Sampling of the neutral species followed by electron impact ionization of the neutrals after they have been extracted from the plasma. Both approaches have advantages and disadvantages and the question as to which approach is most useful is usually answered by a careful consideration of the plasma process under study.

Direct ion sampling requires a somewhat more complex experimental configuration in that the ion beam, extracted through a sampling orifice located in a wall of the discharge chamber, must be focussed into the mass spectrometer. If high mass resolution is not required, quadrupole mass filters are particularly well-suited to this application primarily because these instruments require only low energy ions (~ 10 eV) and can operate without serious problems if the ion beam has a significant energy spread. Magnetic sector instruments require that either the plasma itself or the drift tube of the mass spectrometer can be operated substantially away from ground potential in order to obtain the large and often variable ion kinetic energy required by these instruments. In many ion sampling experiments, the objective is to sample accurately the ion density in the bulk plasma. Ion-molecule collisions in the sheath region and downstream from the sampling orifice distort this measurement and care must be taken to consider these effects[33-35]. It should be mentioned, however, that for those situations in which plasma-surface interactions are the primary interest, the collisional processes in the sheath region are no longer a distortion of the measurement but must be included in order to obtain a true picture of the ionic flux incident on surfaces. One method of determining the importance of collisional processes in the ion extraction is to place an energy spectrometer between the sampling orifice and the mass spectrometer[29]. Ions coming directly from the bulk plasma will have an energy characteristic of the plasma potential; whereas, ions formed in the sheath or downstream from the sampling orifice will have a lower energy. Retarding potential techniques can also be used to obtain similar insight into this problem[36-38] and although energy distributions measured with retarding potential systems tend to be less precise than those obtained with deflection energy spectrometers, the retarding potential approach is compatible with line-of-sight sampling of the neutral species.

Neutral species are best sampled using molecular beam methods in which the neutral beam is modulated between the sampling orifice and the ionization chamber of the mass spectrometer and only the modulated component of the mass spectrometer output is recorded. This approach enables all neutral species, including radicals[39,40], to be detected with a comparable sensitivity. If modulation techniques are not used, the sensitivity for detecting condensible or reactive species is much less than for non-condensible, non-reactive neutral molecules because of the much larger effective pumping speed for the former in the mass spectrometer chamber. However, the ease of installation of non-line-of-sight non-modulated neutral sampling apparatus has prevailed in many cases and useful information about plasma-chemical processes can be obtained in this way.

There are several instances in which direct ion sampling has been used in plasma polymerization studies[37,41–45], but as of this time relatively little ion sampling has been carried out in plasma etching systems[46]. The ability to sample neutral species is usually contained in ion sampling equipment and consequently much neutral sampling work is contained in the ion sampling publications. In addition, there has been considerable neutral mass spectrometry in the experimentally simpler non-line-of-sight configuration.

Optical spectroscopy provides another convenient way of identifying species in a glow discharge. Whereas mass spectrometry by definition is limited to sampling species arriving at the boundary of a plasma, optical spectroscopy can be used to determine spatial concentration profiles[47,48] of the species in the bulk plasma. Probably the greatest advantage of optical spectroscopy for plasma diagnostics is the fact that it can be implemented without modification of the vacuum system provided appropriate windows are available in the plasma reactor.

There are numerous optical spectroscopic methods which could be used for this purpose, but only emission spectroscopy has been used to an appreciable extent in polymerizing[49–51] or etching plasmas[52–58]. Emission spectroscopy has a high sensitivity and is easily implemented but suffers from the disadvantage that the sensitivity is very much dependent on the plasma parameters through N_e and f(E). Consequently, it is difficult to interpret data taken when plasma parameters are varied. This difficulty can be overcome by using the actinometer approach mentioned earlier. Absorption spectroscopy, on the other hand, does not suffer this disadvantage but does not have good sensitivity for detecting molecular species at the low concentrations encountered in etching and polymerizing plasmas. Atomic absorption has good sensitivity but the species of primary interest (e.g., F atoms) absorb in the experimentally inconvenient vacuum ultraviolet region of the optical spectrum. It is anticipated that the ongoing development of stable tunable laser light sources will significantly enhance the use of absorption spectroscopy and stimulated emission spectroscopies in plasma diagnostics. Preliminary results from IR gas phase emission spectroscopy of fluorocarbon discharges look very promising as another diagnostic approach[59] especially well suited to monitor molecular species.

2.2 Characterization of Plasma-Surface Interactions

2.2.1 Etch Rate or Polymerization Rate Measurements

It is clear that a very important measurement both from a technological and mechanistic point of view is the rate at which a surface is etched or the rate at which a polymer film is deposited in a plasma. Essentially all studies of etching or polymerizing plasmas include etching or polymerization rate measurements. Many of these measurements have required that the substrate be removed from the plasma system. Often the substrates are carefully weighed before and after exposure to the plasma and often a profilometer is used to measure the step height across a boundary between a surface exposed to the plasma and a surface shielded from the plasma by a physical obstruction. It is apparent that the requirement that the substrates be removed from the plasma system is a serious experimental limitation.

There are two methods which are used quite regularly to measure etching or polymerization rates dynamically in a plasma environment. One method which is applicable to optically transparent layers on a reflective substrate (e.g., SiO_2 on Si) involves monitoring the optical reflection of a laser beam directed on the surface. As the thickness of the transparent film changes, the amplitude of the reflected beam varies periodically as reflected light from the reflective substrate constructively and destructively interferes with light reflected from the outer surface of the transparent film. A second method involves the use of quartz crystal microbalances[60]. If the process of interest is etching, the material to be etched must be deposited onto the quartz crystals prior to their installation in the vacuum system. These devices are capable of excellent sensitivity (<1 Å/min) and can be operated in a glow discharge environment. An example of the kind of information that can be provided by this dynamic measurement of the etch rate is shown in Fig. 2.5[61]. In this example an oxygen ion beam was directed onto a quartz crystal microbalance coated with silicon on top of which were a few angstroms of carbon. The plot of Fig. 2.5 shows the oxidative removal of carbon (weight loss), the oxidation

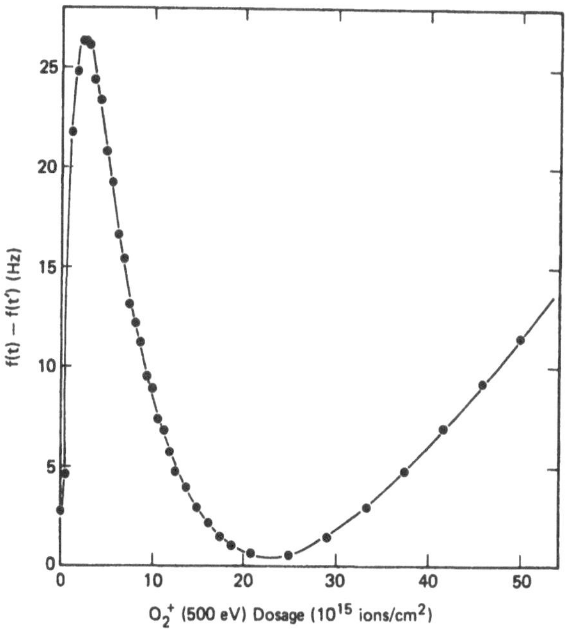

Fig. 2.5. Quartz crystal monitor response when a silicon surface, previously exposed to a high dosage of CF_3^+ ions, is subjected to 500 eV O_2^+ ion bombardment. The regions of this curve are dominated by the following phenomena:
$0–3 \times 10^{15}$ ions/cm² — removal of carbon from Si surface
$3–20 \times 10^{15}$ ions/cm² — oxidation of Si surface
$>40 \times 10^{15}$ — steady state sputtering condition has been reached

of silicon (weight gain) and the steady state sputter-etching of the oxidized surface (weight loss).

Finally, it should be mentioned that, in the case of etching plasmas, any of the mass spectrometric or optical spectroscopic methods described in Sect. 2.1.4 could in principle be calibrated to measure the etch rate simply by monitoring the etch product. This approach has been used for end-point detection in the etch process[52,57,62-64]. These methods may also be applicable to polymerizing plasmas where interesting correlations between the concentration of certain gas phase species and polymerization rate have been noted[45,59,65,66].

2.2.2 Surface and Thin Film Characterization

A very common and useful approach to studying the plasma polymerization process is the careful characterization of the polymer films produced. A specific property of the films is then measured as a function of one or more of the plasma parameters and mechanistic explanations are then derived from such a study. Some of the properties of plasma-polymerized thin films which have been measured include electrical conductivity, tunneling phenomena and photoconductivity, capacitance, optical constants, structure (IR absorption and ESCA), surface tension, free radical density (ESR), surface topography and reverse osmosis characteristics. So far relatively few of these measurements were made with the objective of determining mechanisms of plasma polymerization. The motivation in most instances was a specific application of the thin films. Considerable emphasis on correlations between mass spectroscopy in polymerizing plasmas and ESCA on polymer films with plasma polymerization mechanisms will be given later in this chapter based on recent work done in this laboratory.

Some of the modern surface analysis methods have been used to study the surface chemistry of both plasma etched surfaces and plasma polymerized thin films. Much of this work has involved exposure of these surfaces to air as the samples are transferred from the plasma system to the surface analysis system. However, in a few cases surface analysis has been performed in the plasma system after the plasma gas has been pumped away. This work will be discussed in more detail later in this chapter.

3 Mechanisms in Plasma Etching

In the area of microstructure fabrication it is being recognized that the etching undercut inherent in wet chemical processing poses a serious limitation on the minimum achievable feature size. Consequently, much interest is being directed towards dry etching processes such as plasma etching (directional etching in a chemically active plasma) and ion milling (directional physical momentum transfer sputtering). The redeposition problem[67,68] and the overall lack of selectivity characteristic of ion milling can be overcome in some situations by plasma etching. A major obstacle to the widespread utilization of plasma etching in manufacturing processes is the complexity of the molecular gas glow discharges and of the plasma-surface chemistry which determines the overall etching behavior. The multi-dimensional parameter space associated with these plasmas, illustrated earlier in Fig. 2.1, further complicates the implementation of this technology. Within the last two or three years there has been considerable effort devoted to obtaining a more basic understanding of the fundamental mechanisms which determine etching behavior and in this section the results of some of this work will be summarized. It has been recognized that the complexity of the plasma makes it very difficult to unravel the microscopic details of the important plasma-surface interactions, and consequently, the approach that has been taken is to generate beams of the key species in a plasma and study the interactions of these beams with clean surfaces under carefully controlled conditions. H. F. Winters in a following chapter will describe the basic particle-solid interactions which are important in low pressure plasmas and we will attempt to explain various generally observed plasma phenomena in terms of these basic concepts.

It is useful to clarify the terminology which has evolved with the plasma processing technology. Some factors which are useful in categorizing plasma processes (not including plasma deposition methods) are: a) the nature of the discharge gas, b) the extent to which the surface of interest is subjected to energetic particle bombardment, and c) the volatility of the gas-surface reaction product. These factors are used to classify the various plasma processes in Table 1. In this chapter the term plasma etching will be used more generally to include all processes leading to a volatile gas-surface reaction product and the role of ion bombardment in this process will be considered in some detail.

Very little will be said here concerning the equipment aspects of plasma etching. There are three basic types of equipment which have been used: a) barrel systems, b) planar systems, and c) systems in which the wafers are located downstream from the plasma[69] to be referred to in this paper as downstream etching systems. These plasma etching configurations are shown schematically in Fig. 3.1. Often the barrel systems are used with a perforated metal tube called an etch tunnel[70] which is shown in Fig. 3.1a and b. The purpose of the etch tunnel is to protect the wafers from the energetic ion and electron bombardment to which wafers immersed directly

Table 1. Classification of plasma processes

Type of Discharge			
Extent of Ion Bombardment	Inert Gas Discharge	Reactive Gas Discharge	
		Involatile Gas-Surface Product	Volatile Gas-Surface Product
Minimal (Sample on Ground Plane)	Plasma Cleaning	Plasma Anodization (Oxygen plasma)	Plasma Etching Plasma Ashing (Oxygen plasma)
Extensive (Sample on Target Electrode)	Sputter Etching	Reactive Sputter Etching	Reactive Sputter Etching Reactive Ion Etching

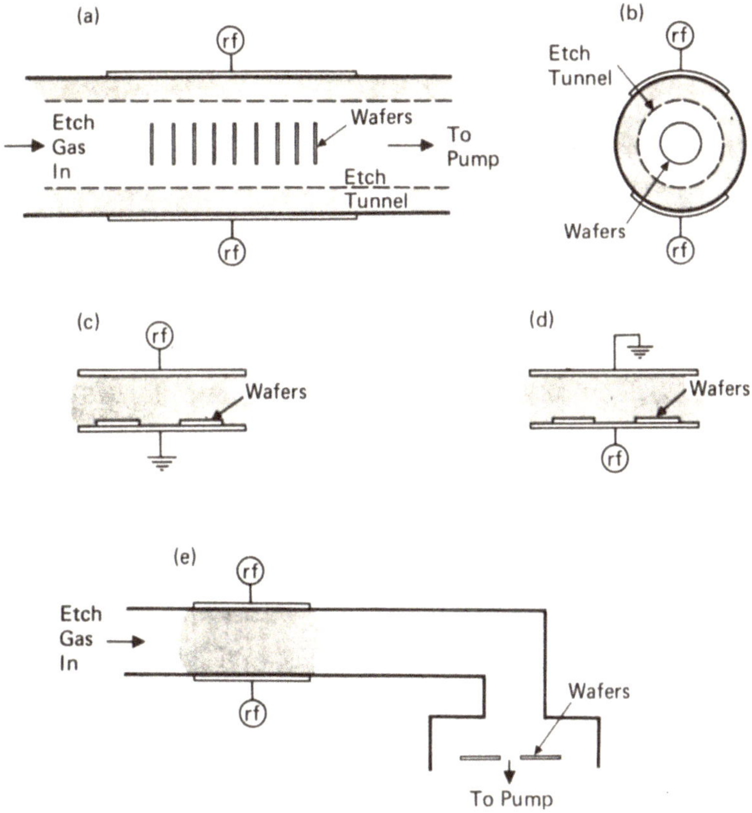

Fig. 3.1. Plasma etching systems. **a** and **b** Barrel system with etch tunnel. **c** Planar system. **d** Planar system (reactive ion etching or reactive sputter etching mode). **e** Downstream system

in the plasma are subjected. The two modes in which planar systems are used are shown in Fig. 3.1c and d where the wafers are mounted either on the grounded electrode[54, 71)] or on the powered electrode[72–74)].

In all these configurations, a major role of the plasma is the generation of chemically active etching species by dissociation of the often relatively inert etching gas. A second important role of the plasma, which is less widely recognized, is to provide energetic radiation (primarily ions and electrons) incident on the surface of interest. In a subsequent chapter, Winters will describe how many gas-surface chemical reactions which do not spontaneously take place in the absence of energetic particle bombardment can be made to proceed rapidly by exposing the surface to a flux of energetic particles. In general, a gas-surface reaction will proceed more rapidly when the surface is subjected to energetic radiation. It is apparent that this radiation-enhanced chemistry cannot take place in the downstream etching system (Fig. 3.1e) nor would one expect it to be important in the barrel system with the etch tunnel in place (Fig. 3.1a and b). However, the ion-enhanced chemistry will be maximized in the planar system with the wafers on the powered electrode as in reactive ion etching (Fig. 3.1d). The extent to which ion-enhanced chemistry participates in the etching reaction determines the etching directionality or anisotropy. This will be discussed later in this chapter.

3.1 Chemical Aspects of Plasma Etching

3.1.1 Role of Oxygen

Oxygen was one of the first gases to be used with fluorocarbon etch gases in the etching of silicon[75]. The etch rate of Si has been found to increase substantially in many etching systems[55,56,72,76,77] when oxygen is added to the CF_4 etch gas and there has been some speculation[76] concerning the chemistry responsible for this etch rate enhancement. Independently, it has been demonstrated that CF_4 does not react with atomic oxygen[78,79] and therefore the chemistry responsible for the Si etch rate enhancement must be a reaction between oxygen and the species formed by the dissociation of the CF_4 molecule. The fact that large quantities of CO, CO_2, and COF_2 are seen in the effluent gas (see Table 2 for a tabulation of the effluent gas mass spectra) when oxygen is added to a CF_4 glow discharge shows conclusively that oxidation of carbon takes place. Even though direct oxidation of CF_4 does not occur, there are various ways in which this oxidation can be accomplished[56], such as the gas phase oxidation of fluorocarbon radicals or the oxidation of carbon or fluorocarbon polymer deposited on surfaces in the plasma etching system. Regardless of how the oxidation of carbon is accomplished, the net result is that fluorine has been made available for etching because the carbon entered the etching system as CF_4.

Harshbarger et al[53] showed that, as oxygen was added to a CF_4 discharge, the F and CO optical emission from the discharge increased roughly in proportion to the increase in the etch rate showing that the etch rate increase is due to an increase in the F atom concentration. Mogab et al.[56] using emission spectroscopy, mass spectroscopy, infrared absorption spectroscopy and titration methods have studied the CF_4—O_2 system in detail and confirm the conclusion that the addition of oxygen to a CF_4 plasma increases the F atom concentration. They point out that in the case of silicon etching there is not a direct proportionality between the Si etch rate and the F atom concentration because of the competition between F atoms and O atoms for active sites on the Si surface. They find that for SiO_2, where the surface is already saturated with oxygen, the SiO_2 etch rate is proportional to the F atom concentration.

Whereas it is clear that the addition of oxygen to a CF_4 plasma increases the F atom concentration, the detailed chemistry responsible for this increase is not clear. The extent to which the oxidation of carbon or fluorocarbon radicals takes place in the gas phase or on the surfaces is not known and inevitably will depend on the plasma parameters. It is not known to what extent the increase in the F atom concentration can be attributed to a new source of F atoms (i.e., oxidation of fluorocarbon radicals) versus a longer lifetime for existing F atoms (i.e., by reducing the loss rate of F atoms caused by recombination processes leading to reformation of CF_4). Possibly the simplest way to envisage the role of oxygen is to consider the effluent gas[80]. Since all the fluorine enters the system as CF_4, it etching

Table 2. Tabulation of Effluent Gas Mass Spectra

Mass Number (amu)	Ion	Probable Parent	Si Target CF_4 Gas	Si Target CF_4 – 18% H_2 Gas	Si Target CF_4 – 18% O_2 Gas	SiO_2 Target CF_4 Gas	Discharge Off CF_4 Gas
1	H^+		0.063	1.86	0.039	0.200	0.011
2	H_2^+	H_2	0.008	0.252	0.006	0.023	0.009
16	O^+		0.25	0.74	1.70	1.02	0.015
18	H_2O^+	H_2O	0.155	0.309	0.151	0.562	0.129
20	HF^+	HF	0.204	12.6	0.692	1.05	0.026
28	CO^+	CO	1.91	2.69	21.4	16.6	0.513
32	O_2^+	O_2	(a)	(a)	6.61	2.29	0.155
38	F_2^+	CF_4, F_2	0.056	0.068	0.093	0.107	0.076
44	CO_2^+	CO_2	0.035	0.129	7.94	7.24	0.007
47	COF^+	COF_2	(b)	(b)	4.68	4.27	<0.001
51	CHF_2^+	CHF_3	0.513	2.51	(c)	(c)	(c)
66	COF_2^+	COF_2	0.191	0.162	2.04	2.24	<0.001
69	CF_3^+	CF_4	100	100	100	100	100
85	SiF_3^+	SiF_4	14.8	11.7	15.8	15.1	<0.001
100	$C_2F_4^+$		0.257	0.282	0.013	0.009	<0.001
119	$C_2F_5^+$	C_2F_6	4.17	0.589	0.191	0.100	0.001
131	$C_3F_5^+$		0.418	0.053	0.011	0.010	0.001
169	$C_3F_7^+$	C_3F_8	0.182	0.018	0.01	0.008	<0.001

(a) obscured by $^{13}CF^+$ but <0.06; (b) obscured by SiF^+ but <0.8; (c) obscured by $^{13}CF_2^+$ but <0.1. Spectra were recorded on a quadrupole mass spectrometer. Ionizing electron energy = 70 eV. Fragmentation pattern for CF_4 gas: CF_3^+ = 100, CF_2 = 9.55; CF^+ = 3.31, F^+ = 2.24, C^+ = 1.95, CF_2^{++} = 1.58, CF_3^{++} = 0.65, F_2^+ = 0.07. (Of these fragment peaks, only F_2^+ was influenced by the discharge.) Glow discharge parameters: pressure = 2.7 Pa (20 millitorr), CF_4 flow rate = 8.4 sccm, target electrode area = 182 cm², 13.56 MHz rf power ~200 watts, inter-electrode spacing = 6 cm, vacuum system volume ~100 liters.

is to occur there must be some way of disposing of the carbon. In the absence of oxygen, the possibilities are limited to the formation of higher order fluorocarbons such as C_2F_6 (see Table 2) or the deposition of carbon or more probably a fluorocarbon polymer on surfaces in the system, or pumping out the fluorocarbon radicals in the effluent gas. When oxygen is present, carbon can be removed very easily as CO, CO_2 or COF_2 (see Table 2) and when these species are seen in the effluent gas, one knows for certain that fluorine has been made available for etching.

3.1.2 Fluorine-Deficient Discharges and Etch Rate Selectivity

An important requirement in certain plasma etching process steps is the ability to etch SiO_2 much faster than Si. With a large SiO_2-to-Si etch-rate ratio, it is possible to etch through a SiO_2 film down to Si without seriously etching into the Si substrate. Heinecke[81)] first demonstrated that this can be accomplished in fluorine-deficient (with respect to CF_4) fluorocarbon plasmas. Heinecke created the fluorine-deficiency both by using hydrogen and by using C_3F_8 as the etch gas intestead of CF_4.

The hydrogen was introduced both as an additive gas (CF_4—H_2 mixtures) and intramolecularly by using CHF_3 as the etch gas. This concept has been extended by several groups and impressive SiO_2-to-Si etch-rate ratios have been obtained. Ephrath[82], for example, using CF_4—H_2 mixtures in a reactive-ion etching system has observed ratios of 30:1. An example of this work is shown in Fig. 3.2. Lehmann and Widmer[83, 84] using CHF_3 gas also in a reactive ion etching mode have obtained ratios of 15:1. We have previously shown[85] that mixtures of CF_4 and C_2F_4 behave similarly to CF_2—H_2 mixtures with respect to the SiO_2-to-Si etch-rate ratio.

It is our opinion that the mechanism responsible for the improved etch-rate ratio is the same for hydrogen additions as for additions of fluorine-deficient gas such as C_2F_4. It is generally agreed that the hydrogen scavenges F atoms to form HF (see Table 2) and even though HF in solution is a good etch for SiO_2, we believe it is relatively unreactive as a diatomic gas phase molecule at the concentrations encountered in plasma etching systems[86]. Consequently, we do not expect HF to participate significantly in the etching chemistry.

The mechanism we believe is responsible for the large SiO_2-to-Si etch-rate ratios which have been obtained in fluorine-deficient discharges is based on several experimental observations. First of all, it has been shown that there are several ways in which carbon can be deposited on surfaces exposed to CF_4 plasmas. One way is to subject the surface to bombardment with CF_3^+ ions which are the dominant positive ionic species in a CF_4 plasma. The extent to which this can occur is shown by the Auger spectra in Fig. 3.3. Curve (a) is the Auger spectrum of a clean silicon surface and curve (b) is the Auger spectrum of the same surface after bombardment with 500 eV CF_3^+ ions. Note that the silicon peak at 92 eV is no longer visible after the CF_3^+ bombardment indicating the presence of at least two or three monolayers of carbon. Another way in which carbon can be deposited on surfaces is by dissociative chemisorption of CF_3 [87] or other fluorocarbon radicals.

A second observation bearing on the selective etching mechanism is shown in Fig. 3.4. In this figure the rate of accumulation of carbon caused by CF_3^+ bombardment is plotted as a function of ion dose for Si and oxidized Si surfaces. It can be

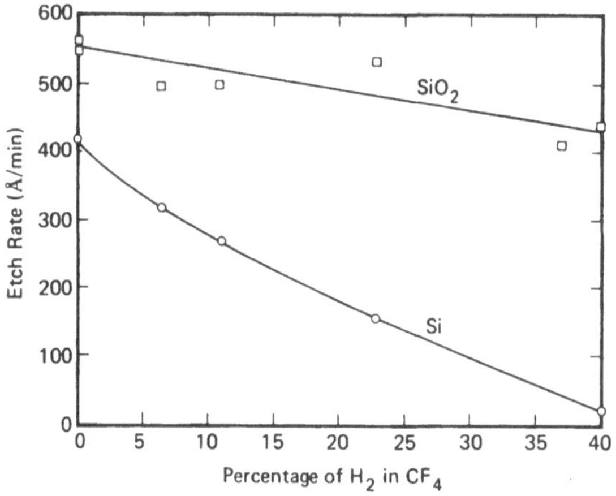

Fig. 3.2. Etch rates of SiO_2 and Si as a function of molecular percentage of hydrogen added to CF_4 in a reactive ion etching system (data from Ref.[82])

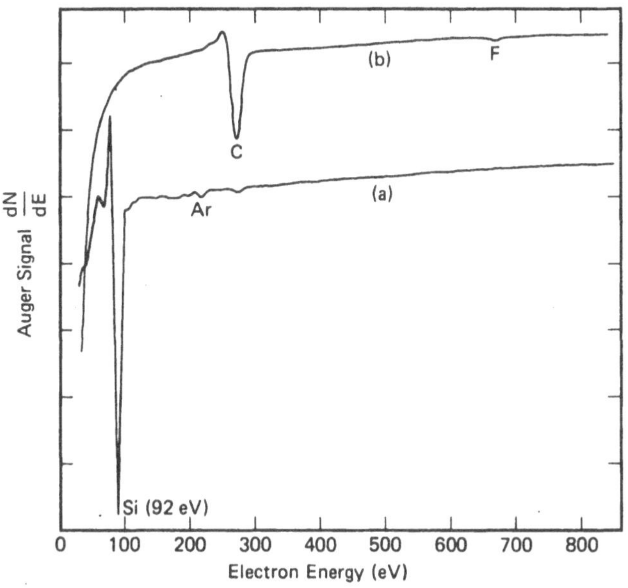

Fig. 3.3. Auger electron spectra from (a) Si surface cleaned by 500 eV Ar^+ ion bombardment and (b) same surface after bombardment with 5×10^{16} CF_3^+ ions/cm^2 (500 eV). (See Ref.[61])

seen that the presence of oxygen in the surface significantly impedes the accumulation of carbon on the surface. Presumably, the oxygen is able to oxidize the carbon and the CO which is formed desorbs from the surface.

The third observation is contained in Table 2 where it can be seen that large amounts of CO, CO_2 and COF_2 are evolved when SiO_2 is etched (SiO_2 target) in a pure CF_4 plasma whereas relatively little of these gases is seen when the target electrode

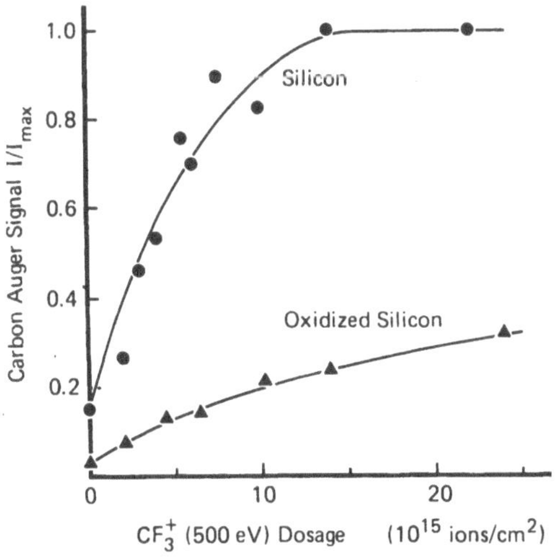

Fig. 3.4. Normalized carbon Auger signal from silicon and oxidized silicon as a function of the dosage of 500 eV CF_3^+ ions incident on these surfaces. (See Ref.[61])

is silicon. That is, the oxygen in the SiO_2 combines with carbon to form CO, CO_2 and COF_2.

Now consider the consequences of making the discharge deficient in fluorine in the light of the above observations. Even in a pure CF_4 discharge, the carbon which is deposited on a silicon surface must be removed, most probably by reforming CF_4. Thus, a fraction of the F which would otherwise be available to form the volatile product of SiF_4 will now be consumed to remove C from the surface. As the discharge is made deficient in fluorine, the ratio of fluorine-to-carbon arriving at the silicon surface will decrease and a larger fraction of the incident fluorine will be consumed by the carbon. Eventually a point will be reached when the carbon consumes all the fluorine and the silicon etching stops. Beyond this point there is insufficient fluorine to volatilize the carbon and fluorocarbon polymer begins to accumulate on the silicon surface. On SiO_2 surfaces, however, the oxygen in the SiO_2 lattice, is able to volatilize the carbon and the incident fluorine is available to etch the silicon in the SiO_2. Eventually, as the discharge is made more and more fluorine-deficient, a point will be reached where fluorocarbon polymer is deposited on the SiO_2 surface as well, but a significantly greater fluorine-deficiency is required for polymerization to occur on SiO_2 than on Si. In summary, in order to obtain large SiO_2-to-Si etch-rate ratios, it is necessary to adjust the fluorine-deficiency of the plasma to the point where there is just enough fluorine incident on the silicon surface to volatilize the carbon which is also incident on the surface. At this point, the oxygen in the SiO_2 lattice will assist in the volatilization of carbon incident on the SiO_2 surface and the fluorine will still be mostly available for etching the Si in the SiO_2.

It should be emphasized that the fluorine-deficiency can be caused by adding a fluorine scavenger such as hydrogen[81,82)], by adding a fluorine-deficient gas such as C_2F_4 [85)], by adding a hydrocarbon gas[88,89)] which contributes carbon and also scavenges fluorine, or by using a solid scavenger of fluorine atoms such as an electrode [46,88)] made of silicon, carbon, teflon, polyethylene, tungsten, etc. It should also be mentioned that this mechanism is not limited to Si and SiO_2 but to any oxide/non-oxide combination of materials[89)]. Nitrogen might be expected to play a similar role in that carbon might be volatilized as cyanogen but efforts to observe carbon-nitrogen species during the etching of nitrides have not been particularly successful[46)].

3.2 The Role of Ion Bombardment in Plasma Etching — Mechanisms for Directional Etching

The importance of ion bombardment in the plasma etching process is most evident in the much larger etch rates which are observed on the target electrode compared to the etch rates measured on the ground plane[72,90,91)]. This difference cannot be easily attributed to physical sputtering although this possibility has not been eliminated and a thesis for a chemically enhanced sputtering process has been presented[92)]. The situation is illustrated in Fig. 3.5 where the silicon etch rate is plotted as a function of bias voltage applied to the wafer in both CF_4 and Ar discharges. It should be mentioned that in this measurement the application of the bias voltage did not change the discharge conditions as judged by the etch rate of a second unbiased wafer. If the sputter yield of CF_3^+ ions is comparable to the sputter

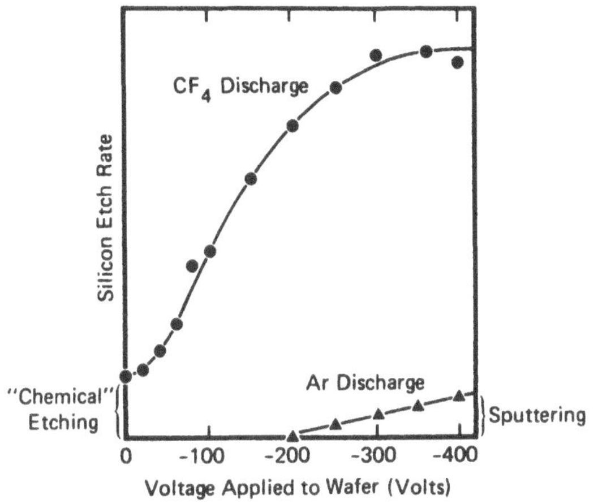

Fig. 3.5. Silicon etch rate as measured with a quartz crystal microbalance as a function of the bias voltage applied to the silicon surface in CF_4 and Ar glow discharges. The discharge intensity was not significantly influenced by the application of the negative voltage to the silicon surface

yield of Ar^+ ions, and some data has been presented which indicates this to be the case[61)], then the etch rate of the biased wafer in a CF_4 plasma cannot be accounted for by superimposing a sputtering rate onto the "chemical" etch rate observed with no applied bias voltage.

The possibility of any "chemical" contribution from the ions can be eliminated in the case of silicon etching with CF_3^+ ions simply by noting that there is not enough fluorine per ion to chemically remove even one silicon atom as SiF_4 and in addition this ignores the problem of the carbon which also arrives at the surface. The observation mentioned earlier involving CF_3^+ ion beam bombardment of silicon in which the silicon surface is soon obscured by a thin carbon layer[61)] emphasizes the fact that CF_3^+ ions alone will not etch Si chemically (by forming SiF_4) at a significant rate.

The process which we believe is responsible for the large increase in the etch rate caused by energetic ion bombardment is the ion-assisted gas-surface reaction[26)] mentioned earlier in this chapter and discussed in detail in a following chapter by Winters. The role of the ions is believed to be primarily physical and not chemical in that ion bombardment of a surface increases the reaction probability for neutral F atoms to react with the surface. The details of how the ion bombardment increases the reaction probability are not clear as yet. Winters[87)] has pointed out that the etching process consists of a series of sequential processes; adsorption, product formation and product desorption. Any of these steps can be the rate limiting step in the etching process, or if any of these steps does not take place, the surface will not be etched. It is not unreasonable to expect that ion bombardment could enhance the rate at which any or all of these three steps proceed. Consequently, depending on which step is rate limiting, the specific mechanism by which ion bombardment enhances the gas-surface reaction rate can be expected to be different for different chemical systems. This subject has been discussed in more detail elsewhere[26)].

It is our current opinion that all etching anisotropy or directionality is caused by energetic particle bombardment (primarily ions but in some cases, electrons). Other factors such as the degree of fluorine deficiency in the discharge can influence the etching anisotropy substantially but these other factors by themselves cannot cause directional etching. In most etching situations, the topographical structure of the substrate and masking layers is very small in comparison to the sheath thickness at the substrate-plasma boundary. Consequently, ions are normally incident on the bottom surfaces of etched features but to a first order approximation the sidewalls or edges of etched features are not subjected to energetic ion bombardment. As was discussed earlier in this chapter, the energy of the ions incident on the etched surface is determined by the potential of the surface relative to the potential of the plasma. In reactive ion etching systems, ion energies of the order of several hundreds of eV can be expected. However, even when the surface to be etched is placed on the ground plane, high energy ions are sometimes encountered because of a large plasma potential. As far as electrons are concerned, Winters in a following chapter will describe how secondary electron emission from the target electrode is a major source of high energy electrons on the ground plane. A similar mechanism applies to negative ions which can be formed on the target surface[93, 94]. The trajectory of these energetic negative species is determined by the orientation of the target electrode in that these particles are accelerated away from the target by the electric field in the target-plasma sbeath. If the target is parallel to the ground plane, as is almost always the case, the energetic negative particles will be incident normally onto the substrates. However, if the target is not parallel to the substrate surface, non-normal incidence of energetic negative particles will occur.

The important point to recognize is that the etch rate of surfaces subjected to energetic particle bombardment (bottom surfaces) will be larger than the etch rate of surfaces not subjected to this bombardment (sidewalls) because of the ion-assisted (or electron-assisted) gas-surface chemistry. The relationship between the shape of an etched profile and the dependence of the etch rate on ion bombardment is shown in

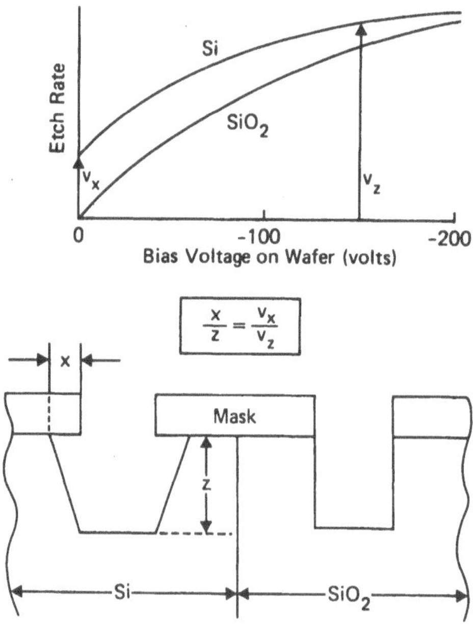

Fig. 3.6. Illustrative figure to show the relationship between the shape of the etched wall profile and the dependence of the etch rate on the wafer potential (see text for discussion)

Fig. 3.6 for hypothetical Si and SiO_2 samples subjected to 150 eV positive ion bombardment. The —150 volt bias on the sample determines the etch rate of the bottom of the etched feature (V_z) whereas the lateral etch rate can be approximated by the zero bias etch rate (V_x). This latter quantity is dependent on the material being etched and is showed as zero for SiO_2 in Fig. 2.6. Consequently, the etched feature in the SiO_2 has vertical sidewalls. Since V_x is not zero for the Si, the sidewalls are not vertical.

For systems in which V_x is not zero, the shape of the etched profile can be controlled to a certain extent by adjusting the stoichiometry of the discharge. This is illustrated in Fig. 3-7 using an idealized Si sample. As hydrogen is added to a CF_4 discharge the etch rate of Si will decrease as we have seen earlier in Fig. 3.2. However, the etch rate will stop on surfaces not subjected to ion bombardment (point A in Fig. 3.7) before etching stops on surfaces which are exposed to energetic ion bombardment. This means that the lateral etch rate has been eliminated and features with vertical sidewalls can be etched if an etch gas mixture of CF_4 — 10% H_2 is used in this example.

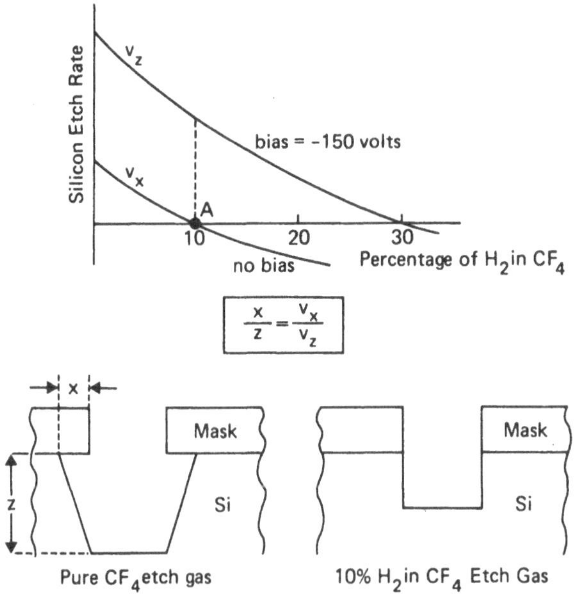

Fig. 3.7. Illustrative plot to demonstrate the way in which the shape of the etched wall profile can be influenced by introducing a fluorine-deficiency in the discharge, in this example by adding hydrogen (see text for discussion)

3.3 Consumption of the Active Species by the Etching Process — The Loading Effect

A characteristic of the plasma etching process which is generally observed is that the etch rate of a sample decreases as the area of the sample exposed to the plasma increases[54, 73, 76, 95]. This dependence of etch rate on batch size is referred to as the "loading effect" and an example is shown in Fig. 3.8. The reason for the loading effect is simply that the etching process consumes a significant fraction of the

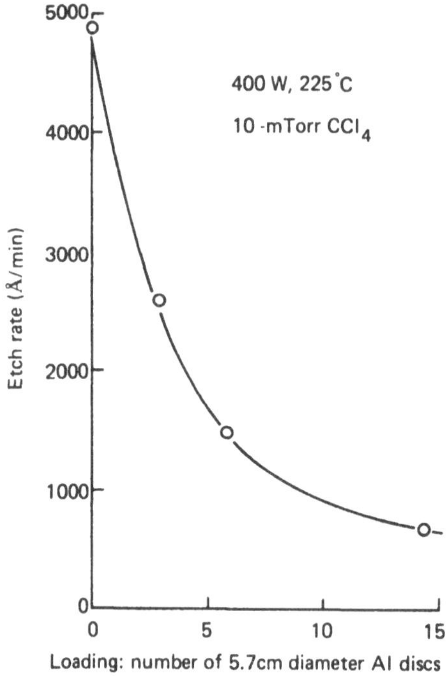

Fig. 3.8. The loading effect (etch rate versus batch size) observed during the etching of Al in a CCl_4 discharge in a reactive ion etching configuration (data from Ref.[87])

available etching species and when the area of etchable material is increased the partial pressure of the etching species (and therefore, the etch rate) decreases. "Available" species in this sense denote the fraction of the etching species consumed by reactions in which the rate of reaction depends upon the density of these species. If processes exist which consume a constant amount of the etching species, independent of concentration, these species are not "available" for etching. Mogab[54] has discussed the loading effect in detail and has shown that a useful way of displaying loading data is to plot the reciprocal of the etch rate versus the area of etchable material. If the etch rate is proportional to the density of etching species, the fraction of the "available" etching species which is consumed by the etching process can be determined directly from such a plot. This concept is illustrated in Fig. 3.9 using the data of Mogab[54].

Any method of observing the concentration of the etching species can be used to monitor the loading effect. Both Harshbarger et al.[53] and Mogab[54] have used the F optical emission from the plasma to monitor the F atom density in the presence of varying areas of silicon in the system. These authors point out how this emission signal can be used as a process monitor. In addition, the consequences of loading have been observed indirectly using mass spectrometry[46]. F atoms are difficult to observe mass spectrometrically because of their high reactivity but excessive consumption of fluorine causes the gas phase species to be deficient in fluorine and significant quantities of unsaturated fluorocarbons are seen when large areas of silicon are present in the etching system.

Clearly, the existence of a loading effect is not desirable in a manufacturing process. However, it is not obvious that the effect can be eliminated without

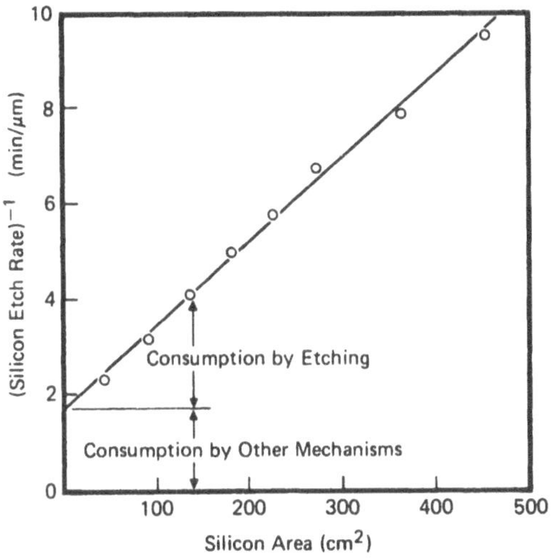

Fig. 3.9. Reciprocal of the silicon etch rate versus silicon area illustrating the relative consumption of the F atoms by the etching process (data from Ref.[54])

introducing other problems. Mogab[54] has pointed out that in order to eliminate loading, it is necessary to have short-lived etching species. That is, the consumption of active species by non-etching processes must be much larger than the species consumption by etching. Sputter-etching is an example of such a situation in which the etching species (energetic positive ions) are very short-lived and consequently there is no loading effect. Another rather subtle effect of the loading phenomenon has been mentioned by Mogab[54]; namely, that unless care is taken to maintain a constant concentration of etching species, activation energies derived from the temperature dependence of the etch rate will be in error.

3.4 Simplifying Concepts

Several concepts have evolved in the study of plasma etching mechanisms which can sometimes assist in keeping the various interacting phenomena in perspective. One approach is to consider the plasma etching system as a "black-box"[80] and deduce the chemistry taking place in the black-box by monitoring the input gases and the effluent gases. The techniques which have been used for effluent gas monitoring were discussed earlier in this chapter. The black-box approach consists simply of requiring that the amount of C & F leaving the system in the effluent is equivalent to that entering the system as etchant gas. For example, consider the CF_4 plasma etching of Si or SiO_2. Typical gases found in the effluent are (see Table 2) CF_4, C_2F_6, SiF_4, CO, CO_2, COF_2, HF, and F_2. If we assume negligible fluorocarbon polymer deposition on internal surfaces of the vacuum system, we can obtain the following expression by requiring the conservation of fluorine and carbon:

$$Q(SiF_4) = (^1/_2)\,Q(C_2F_6) - (^1/_2)\,Q(F_2) + (^1/_2)\,Q(COF_2) - (^1/_4)\,Q(HF) + Q(CO) + Q(CO_2)$$

where the Q's are the number of molecules per second leaving the etching system in the effluent gas. If other gases are observed, the expression above can be easily modified to include the contributions of these additional gases. Many of the concepts that we have discussed earlier in this chapter are contained in this expression. For example, the role of oxygen in oxidizing carbon thus making fluorine atoms available for etching is implicit in this equation. The correlation between the CO evolution and the etch rate[53, 56] and between the COF_2 concentrations and the etch rate[76] are also contained in this expression. The role of hydrogen in consuming fluorine to form HF thus decreasing the etch rate and the importance of C_2F_6 in oxygen-free systems is also suggested. This expression can also be used to show why SiO_2 does not cause loading to the same extent as Si. The reader is referred to a detailed discussion of these concepts elsewhere[80].

A second simplifying concept is to characterize the discharge by the relative stoichiometry (fluorine-to-carbon ratio[85] or F/C) of the *active species*. In determining this F/C ratio, gases which do not react significantly at the surface being etched or which are thought not to participate significantly in polymerization (e.g., CF_4, CO, CO_2, COF_2, F_2, etc.) are ignored whereas species which do react (e.g., CF_3, CF_3^+, CF_2, CF_2^+, CF, CF^+, F, C, etc.) are included. That is, the F/C ratio of the active species is determined by the F/C ratio of the etch gas modified by processes which result in fluorine or carbon being converted into unreactive species. For example, the F/C ratio of the active species generated by the dissociation of CF_4 is four. However, the etching of Si will consume F atoms forming relatively unreactive SiF_4, thus decreasing the F/C ratio of the active etching species. A similar effect is achieved by adding hydrogen or an unsaturated fluorocarbon gas or as discussed earlier in this chapter. If oxygen is added to the system, active carbon is converted to inactive CO or CO_2 thus increasing the F/C ratio. As the F/C is decreased the SiO_2-to-Si etch-rate ratio increases until eventually polymerization begins to dominate over etching. The boundary between polymerization and etching and its dependence on the bias applied to the surface under consideration can be represented qualitatively using the F/C ratio. This is shown in Fig. 3.10 and it should be emphasized that shape and location of this boundary is not well established and will depend on many system parameters. It was implied earlier in Fig. 3.7 that ion bombardment tends to retard the onset of polymerization, and quite often in plasma etching systems one has etching of the target electrode and polymerization on grounded surfaces[46]. This condition is depicted in Fig. 3.10 for $2 < F/C \leqq 3$ and we will discuss later in this chapter how this operating condition can be exploited to form interesting metal-containing polymeric thin films[65, 66].

This F/C ratio concept suggests that the molecular nature of the etch gas as injected into the system is not particularly important and that the etching chemistry can be correlated to the atomic composition of the discharge gases. This has been found to be true for systems with long residence times (τ) for the gases in the discharge (i e., low flow rates). Very similar etching behavior has been noted for Si with COF_2 and C_2F_4—O_2 (1:1) etch gases[85]. Even a pure oxygen discharge with a polytetrafluoroethylene excitation electrode gave similar effluent gas mass spectra when compared with the COF_2 and C_2F_4—O_2 (1:1) discharges[85]. This concept will not be valid for very short residence time systems, but it is suspected that it is applicable to many of the configurations used in plasma etching.

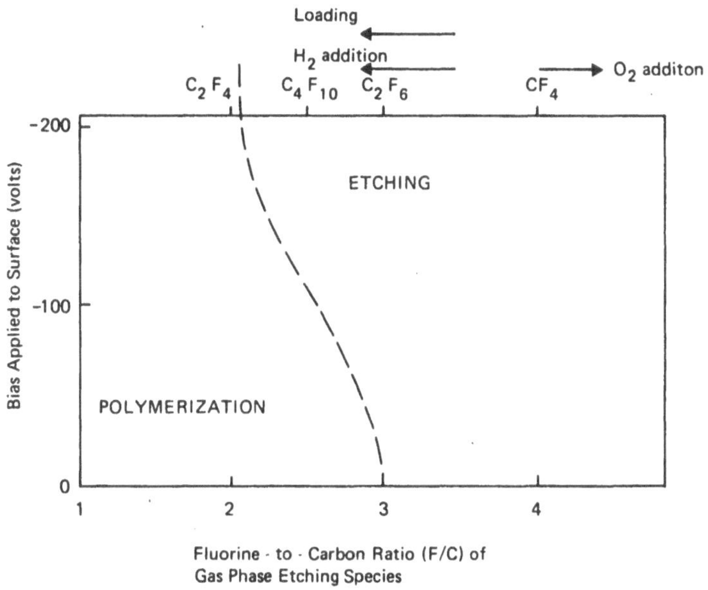

Fig. 3.10. Schematic representation of the boundary region between etching and polymerization as influenced by the fluorine-to-carbon ratio of the etching species and the bias applied to the surface of interest (see text for discussion)

A third simplyfying concept is the utilization factor introduced by Chapman and Minkiewicz[32)] which is, in the case of CF_4 etching of Si for example, simply the ratio of the rate at which fluorine is consumed in the etching process to the rate at which fluorine is injected into the system as CF_4. (Utilization factor = $Q_{out}(SiF_4)/Q_{in}(CF_4)$ where Q_{out} and Q_{in} are the flow rates out of and into the etching system respectively.) In any etching process, the utilization factor should be substantially less than unity, say of the order of 0.1 or less.

4 Plasma Polymerization of Fluorocarbons

In recent years, there has been an increasing interest in plasma polymerized films of organic monomers[96–99)] (the term monomer is used in this context to specify the low molecular weight starting material rather than the true precursor to polymerization), because of their interesting chemical, mechanical, thermal, optical and electrical properties, all of which can be quite different from and often superior to the conventionally polymerized films. The polymers formed by the discharge technique on suitable substrates may find wide ranging applications in a variety of technologies and by varying the composition of the gas injected into the plasma, certain physical properties of these mostly dielectric and chemically very stable films can be varied systematically. The possibility of adjusting the refractive index, for example, makes it possible to use these films for optical waveguides[100] and antireflection coatings[101] while a close control of the density of crosslinking in the films suggests utility in, for example, desalination[102] and gas separation applications[103]. Other applications range from orienting layers for liquid crystals[104, 105] to corrosion protection and from protection of semiconductor devices[106] to the use as dielectrics in thin film capacitors[107].

Although to-date the emphasis has been on plasma polymerized films produced from hydrocarbon based systems, this trend in more recent times has swung towards fluorocarbons in an attempt to produce polymers of similar properties to conventionally prepared linear fluoropolymers. However, it will become clear from the account to follow that in many respects plasma polymerized fluorocarbons differ significantly from their linear counterparts. It is to the plasma polymerization of organic monomers containing solely carbon and fluorine therefore that we shall devote our attention in this section with only brief references to hydrocarbon and fluorohydrocarbon polymers[125] for comparison purposes.

As has been pointed out previously in this chapter and elsewhere[108], the ability of fluorocarbons to undergo polymerization in a plasma is related to the fluorine-carbon stoichiometric ratio.[109] Although CF_4 has been observed to polymerize in high power, magnetically confined plasmas[110], it is only for fluorocarbons of $F/C \leqq 2$ that a significant rate of polymerization is observed under normal conditions[108] in the absence of other effects[64-66] such as, etching, hydrogen addition, etc. For this reason along with the well established chemical and thermal stability and high electrical resistance of conventionally prepared polytetrafluoroethylene, the plasma polymerization of tetrafluoroethylene has been the most widely studied system of all fluorocarbons and will form the basis of many of the discussions to follow.

4.1 Gas Phase Species in a Tetrafluoroethylene Plasma

Although both neutral and ionic species have been monitored by 'in situ' mass spectrometric analysis of C_2F_4 plasmas, this discussion will be largely restricted to the neutral chemistry since in general the amount of polymer deposited relative to the amount of monomer fed into this system is too large for ionic spcies to be solely responsible for the process[45]. This is not, however, to say that the ionic species do not play any role in the polymerization process and the importance of the ions in the plasma polymerization of fluorocarbons will become apparent.

The polymer deposition rate of C_2F_4 in a plasma and the neutral mass spectra have been measured as a function of the monomer input flow rate, keeping the pressure and power constant (0.015 Torr, 35 watts)[45]. As is the case with many hydrocarbons[99, 111], the deposition rate has a pronounced maximum in the accessible flow range. At very low flow rates (long residence times), the deposition rate is limited either by the amount of monomer available or by excessive fragmentation of the monomer; whereas at high flow rates, the low deposition rate is probably due to the fact that the monomer molecules and active species are swept from that region of the system where polymerization takes place before undergoing reaction. Unambiguous identification of the neutral species extracted from the plasma was made possible by using a low electron impact ionizing energy (20 eV) in the mass spectrometer which minimizes the degree of fragmentation[45] and allows the use of appearance potentials for further clarification. It was observed in the C_2F_4 plasma that the amplitudes of the signals due to saturated species, and thus the concentrations of these species in the plasma, did not correlate with the polymer deposition, whereas, those associated with unsaturated species did. In fact, the observed intensities of the unsaturated neutrals in plasmas of a large variety of injected fluorocarbons also correlated quite well with the deposition rate. Thus in the case of rapidly polymerizing monomers, such as perfluorobut-2-ene and perfluorocyclobutane, these species are observed with intensities comparable to those found for the C_2F_4 plasma while for monomers which do not readily polymerize, such as CF_4 and C_2F_6, the intensities of these species are relatively low. However, the addition of a small amount of hydrogen to either CF_4, C_2F_6 [45] or C_3F_8 [63] greatly increases their polymerizability and this effect is again manifested in the neutral spectra by a large increase in the intensities of the signals due to unsaturated species, and in particular C_2F_4.

'In situ' mass spectrometric investigations of the C_2F_4 glow discharge have also been carried out by Vasile and Smolinsky[44]. From this work, they have concluded that in contrast to hydrocarbon systems, fluorocarbon plasmas produced a high yield of gaseous products as well as polymer. In the particular reactor and mass spectrometer arrangement employed (20 eV electron impact ionization energy), the major peaks in the neutral spectrum were assigned to a convolution of the fragmentation patterns of the unreacted C_2F_4[44]. No evidence for the effusion of atomic or molecular fluorine from the discharge could be found which suggests that these latter two species are most likely present in a very low steady-state concentration, since their reactivity in the discharge environment must be very high. On the other hand, appearance potential measurements confirmed that difluorocarbene ($:CF_2$) did effuse from the plasma confirming the presence and the 'long-lived' nature of this diradical in the plasma.

4.2 Plasma Polymerization Mechanisms

The following series of reactions, therefore, are proposed[44] to account for the neutral chemistry in the C_2F_4 plasma (for species of four or less carbon atoms).

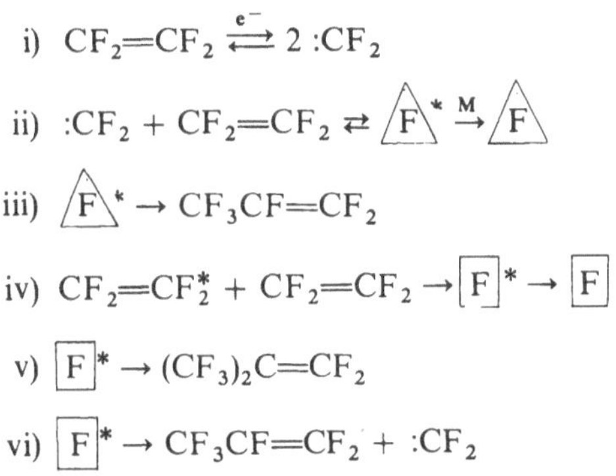

vii) $CF_2{=}CF_2 \xrightarrow{e^-} CF{\equiv}CF + 2F$ (or F_2)

viii) $CF_2{=}CF_2 \xrightarrow{e^-} CF_2{=}CF + F$ (or $1/2\,F_2$)

ix) $CF_3CF{=}CF_2 + F_2 \rightarrow C_3F_8$

x) $CF_2{=}CF_2 + F_2 \rightarrow C_2F_6$

Reactions (i)–(viii) therefore describe the formation of unsaturated species, although reactions (vii) and (viii) are only expected to play a very minor role. Reactions (ix) and (x) produce saturated species which may undergo fragmentation to form $CF_3\cdot$ and $C_2F_5\cdot$ radicals.

Dilks and Kay[65,66] have noted the presence of $(CF_2)_n$ species in the effluents of plasmas excited in several different kinds of fluorocarbons and have proposed that these are the primary precursors to plasma polymerization of fluorocarbons of $F/C \geqq 2$, since their gas phase concentration is found to be directly related to deposition rate for a variety of systems[65,66]. The presence of high molecular weight oligomers in the gas phase of fluorocarbon plasmas has been reported by several workers[44,45] and in particular, in the investigation of a C_2F_4 discharge by Dilks and Kay[65,66] where fragments containing as many as twelve carbon atoms were observed. Since in that work, the analysis was limited by the sensitivity and mass range of the mass spectrometer, this in all likelihood represents a lower limit to the maximum oligomer size[65,66].

Clearly, the gas phase reactions must play an important part in the plasma polymerization of fluorocarbons, in the pressure regime where the gas phase collisional frequency is sufficiently high. The general concensus of opinion that difluorocarbene plays a central role is further supported by investigations involving direct photolysis of C_2F_4 by 1849 Å radiation[112] and trapping of perfluoroalkyl radicals produced in a C_2F_4 discharge on a cold finger maintained at liquid nitrogen temperature[113].

Dilks and Kay[66] have proposed that oligomerization may occur by addition of $:CF_2$ to C_3F_6, C_4F_8, etc. or by a radical addition chain mechanism involving C_2F_4, C_3F_6 and C_4F_8. The primary precursors to polymerization having the general formula $(CF_2)_n$ and including both cyclic alkanes and mono-olefins may, therefore, polymerize in a fashion analogous to conventional polymerization via homogeneous or heterogeneous chain reactions. However, the relatively large concentration of initiators and possible chain terminators in the plasma as well as the low pressure of the system will ensure that the chain length of these initial polymer products are short. Since the final stoichiometry of polymers formed in a C_2F_4 discharge is generally reported to be $\sim C_1F_{1.4}$ [114], this polymerization process involving $(CF_2)_n$ is probably accompanied by a concommitant interaction of photons, ions and metastables with the forming polymer to eliminate fluorine atoms[115], and at higher power loadings, small fragments by ablation[99]. These latter processes, therefore, result in the formation of crosslinks, unsaturated and free radicals in the polymer structure. The unsaturated features and free radical sites may further react with incoming lower molecular weight free radicals ($CF_3\cdot$, $C_2F_5\cdot$, oligomers, etc.) or may remain in the final polymer product.

This proposed mechanism for the fluorocarbon polymer formation is of course not the only mechanism which is conceivable, but is, we believe, the most probable in the light of the available experimental data relating to both the gas phase species and polymer product[66].

4.3 Polymer Characterization

Due to the highly crosslinked insoluble nature of plasma polymerized films in general, the techniques available for their structural characterization have been limited to those for which solid samples can be investigated, although the investigation of pyrolytic decomposition products by gas chromatography also provides useful structural information[116]. Conventional polymer characterization methods, such as differential scanning calorimetry and thermal gravimetric analysis have been found to yield largely featureless spectra, and while infrared spectroscopy has proved to be very useful in the study of hydrocarbon plasma polymers, for fluorocarbon materials only broad, ill-defined bands are discernable[98]. The most successful spectroscopic technique for determining the structure and bonding in plasma polymerized fluorocarbons to-date has proved to be X-ray photoelectron spectroscopy (XPS or ESCA), although the rapidly growing field of solid state nuclear magnetic resonance spectroscopy for the study of polymers may well become a strong contender in the near future[117]. Electron spin resonance spectroscopy usefully provides information with regard to the number and lifetimes of trapped free radicals in the polymers, while water and hydrogen contact angle measurements provide complimentary data to ESCA concerning the outermost surface of the sample. Micro-analytical techniques of course provide direct information relating to the overall elemental composition of the plasma polymerized films.

An investigation of polymers produced both in a tetrafluoroethylene glow discharge and at some distance downstream of the discharge has been reported by Rice and O'Kane[114,118]. Analysis of the polymers was achieved mainly by ESCA although infrared data was also recorded. The surface structure as seen by ESCA was confirmed to be the same as that of the bulk material by comparison with bulk carbon and fluorine analysis. The assignment of particular C_{1s} binding energies was based on both bulk standards and a simple electrostatic model for group electronegativities.

The highly crosslinked nature of the polymer formed in the discharge is immediately apparent from the C_{1s} spectrum which exhibits approximately equal quantities of the four major carbon environments; CF_3, CF_2, CF and carbon with four nearest carbon neighbors. The overall stoichiometry determined from a comparison of the F_{1s} and C_{1s} peak intensities is reported to be $\sim C_1F_{1.4}$. The amount of oxygen functionalities found in these polymers, due to reaction of free radical sites in the polymer with atmospheric oxidants is usually less than 0.1 atomic % and is observed in the ESCA spectrum as a small broad signal in the O_{1s} region[66].

In contrast to the film prepared in the plasma, a film prepared downsteam in the system shows little evidence for crosslinking[114,118] and consists of mainly CF_2 type carbon with a lesser amount of CF_3, CF and C environments, its overall atomic composition being $\sim C_1F_{2.2}$. The spectrum of this film (which incidentally is not unlike that which would be expected to result from radical or ionic polymerization of the $(CF_2)_n$ species discussed earlier) could be dramatically converted to a structure similar to that recorded for the polymer formed in the plasma by subsequent argon ion bombardment of the sample[114]. In addition, the infrared spectrum of the film obtained downsteam resembles that of conventional polytetrafluoroethylene while that of the polymer formed in the plasma is found to be largely featureless.

These results provide extra evidence for the hypothesis of simultaneous chain growth polymerization and fluorine elimination via interaction with energetic species in the formation of plasma polytetrafluorethylene. This is also consistent with the observation that for a plasma of C_2F_4 confined in a closed system the pressure decreases initially to a minimum (polymerization) and then increases (fluorine elimination) producing a non-condensible gas[108].

In an investigation by Clark and co-workers[98] plasma polymers derived from the series (perfluorobenzene, perfluorocyclohexa 1,3 and 1,4 dienes, perfluorocyclohexene and perfluorocyclohexane) have been subjected to ESCA analysis. The importance of unsaturation in the plasma polymerization of fluorocarbons is confirmed by the observation that the deposition rate for these systems increases with the number of double bonds in the starting material. The ESCA spectrum of the polymer derived from cyclo-C_6F_{12} is very similar to that derived from C_2F_4 having the same monomer F/C ratio; however, as the F/C ratio is decreased along the series in replacing CF_2 by CF in the monomer, an increasing proportion of CF type environments is evidenced by the C_{1s} spectrum of the polymer[98]. Although no mass spectrometric analysis has been carried out on these systems, it would seem likely that while polymerizable species like $(CF_2)_n$ predominate in the gas phase of fluorocarbon plasmas where F/C = 2 for the starting monomer, when the F/C ratio is reduced, species such as $(CF)_n$ may become more important. Indeed, the observation that a significant quantity of unsaturated sites are present in plasma polyperfluorobenzene suggests that either difluoroacetylene or perfluorobenzene itself plays the dominant role in polymerization[98].

In general, the concentration of free radicals in plasma polymers produced from fluorocarbon monomers has been found to be somewhat higher than that of other organic compounds with the exception of acetylene and ethylene[108], although the number of free radicals can be reduced to some extent by employing a pulsed discharge[108]. A typical range for the spin concentration for plasma polymers in general is $\sim 10^{18}$–10^{20} spins cm^{-3} [108]. The ESR spectrum of a typical plasma polymerized fluorocarbon consists of a broad singlet with a line width of ~40 gauss[108,119], which is somewhat larger than for polymers produced from other organic monomers (e.g., hydrocarbons = 15–20 gauss). In plasma polytetrafluoroethylene and other plasma polymerized fluorocarbons, this has been attributed[108] to an abundance of polyethylene-type radicals. This is in contrast to the situation found for plasma polymerized fluorohydrocarbons where polyene-type radicals have been proposed[108] to be most abundant presumably due to the possibility of HF elimination which is energetically favored over F_2 elimination both in the gas and solid phases.

4.4 Some Physical Properties of Plasma Polymerized Fluorocarbons

The critical surface tension has been evaluated by means of advancing contact angle measurements for water, various hydrocarbon liquids and for plasma polymerized fluorocarbon films[108,118,119]. The most comprehensive, however, is that of Yasuda and Hsu[108]. The critical surface tension of a given plasma polymer will undoubtedly depend on the degree of crosslinking and unsaturation in the film which in turn depend to some extent on the reactor configuration and plasma operating parameters (power, pressure, flow rate) employed in the polymer synthesis. It is therefore not appropriate to quote absolute values here, but rather discuss them in relative terms. For the majority of fluorocarbon films prepared in a plasma to-date, the advancing contact angle for a given liquid has never been observed to exceed that measured for conventionally polymerized tetrafluoroethylene[108], although monomers containing a relatively large degree of CF_2 and particularly CF_3 groups produce films exhibiting contact angles fairly similar to that of polytetrafluoroethylene. The instances where so called plasma polymerized films show a contact angle greater than that observed for polytetrafluoroethylene are in general those which are formed from monomers of high F/C ratio ($\sim$2) downstream of (or spatially removed from) the discharge region[118] in which case the crosslink density is considerably reduced (i.e., these films resemble conventionally polymerized materials rather than plasma polymers). The contact angle, therefore, generally speaking, decreases with decreasing fluorine content of the polymer[108]. Thus, while tetrafluoroethylene and n-perfluoro-hex-1-ene for example, give polymers which exhibit high contact angles (low critical surface tension or surface free energy) hexafluorobutadiene and perfluorobenzene give polymers which exhibit lower contact angles. The possibility of producing low surface energy coatings of the order of a few tens of Angstroms thick has found considerable utility in soil- and water-proofing of fabrics, for example[120, 121].

Plasma polymerized films are uniform adhere well to substrates and are pinhole-free[110]. However, they are not impervious to small molecules, despite their highly crosslinked structure. Several recent publications have attested to the fact that plasma polymers derived from hydrocarbon based monomers can be effectively used for desalination[102] and gas separation purposes[103]. Furthermore, Diaz, Kanazawa and Kay[122] have shown that in an experiment involving the oxidation and reduction of ferrocene in an electrolytic cell, by means of a platinum electrode coated with several hundred Angstroms of plasma polytetrafluoroethylene, the reaction is rate limited by diffusion of the ferrocene molecules in and out of the polymer film to the electrode surface. This experiment suggests that although the crosslink density is high for the polymers, the relative geometry of the crosslinks seems to be such that 'interchain' spaces remain which are sufficiently large to allow ferrocene to pass through[122].

Since plasma polymerized films have been found particularly useful in the fabrication of thin film capacitors and as insulating layers, a knowledge of their electrical characteristics is desirable. A systematic investigation of the ac-conductivity and capacitance over a wide frequency range (10^{-3}–10^5 Hz) has therefore been conducted by Hetzler and Kay[123] for plasma polytetrafluoroethylene films. The films were studied sandwiched between gold electrodes at a variety of temperatures. At room temperature and below, a plot of the conductivity versus the log of the frequency reveals a power law with a power of unity within the experimental error over a wide range of frequencies (the conductance $\sim 10^{-13}\ \Omega^{-1}\ \mathrm{cm}^{-1}$ at 100 Hz and is thickness independent). Such a power law dependence is generally interpreted as being due to hopping conductivity. Deviation from the power law is observed only at very low frequency with leveling out to the dc-conductivity which was estimated

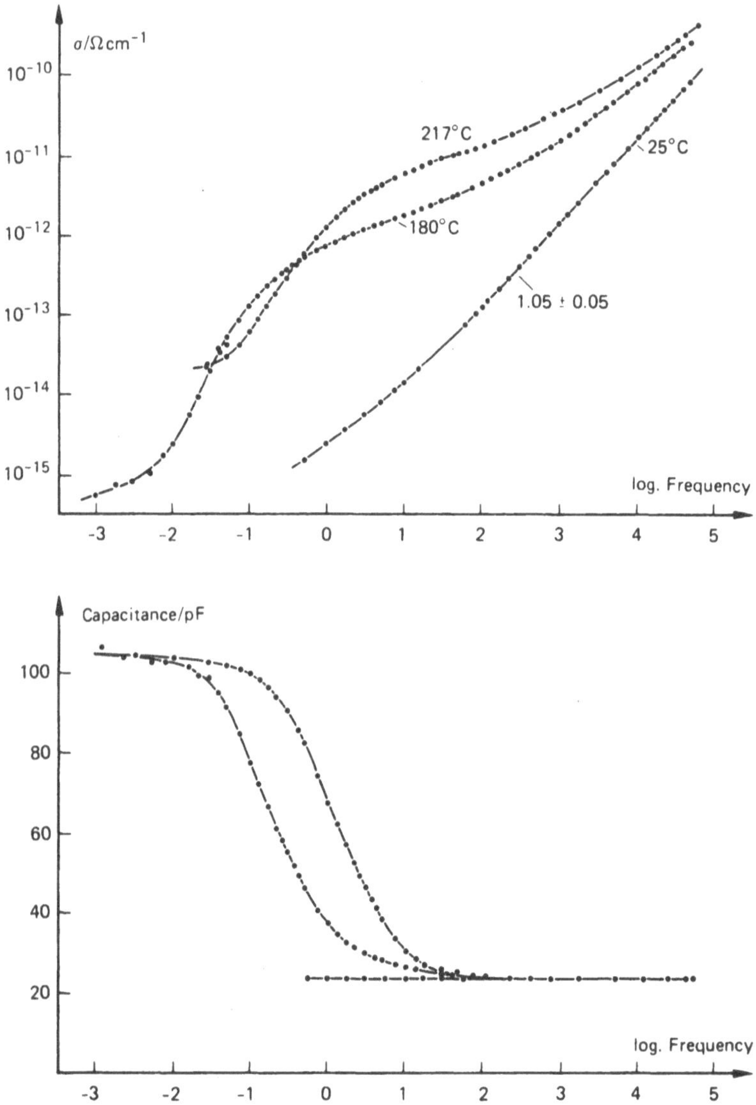

Fig. 4.1. Ac-conductivity and capacitance as a function of frequency for three different temperatures

to be $\sim 10^{-17}\ \Omega^{-1}\ cm^{-1}$. The corresponding capacitance is essentially constant over the entire frequency range (see Fig. 4.1).

At elevated temperature, this behavior is changed drastically. Whereas the high frequency conductivity increases only weakly, large changes are observed at low frequency. After first leveling out at ~1 Hz at 180 °C and ~10 Hz at 217 °C, the conductivity drops several orders of magnitude and then levels out at the final dc-conductivity which could be attained only at 180 °C when measurements were performed at very low frequency. With increasing temperature, the whole curve is shifted to higher conductivity and frequency without changing the general shape. In

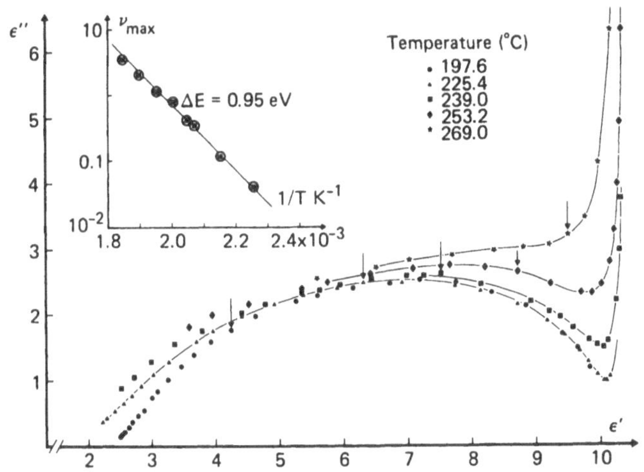

Fig. 4.2. Imaginary part ε'' of the complex dielectric constant versus real part with frequency as a parameter (Cole-Cole plot) at different temperatures. Arrows indicate the frequency of 10 Hz in each case. Insert shows thermal activation energy plot. (See Text)

the frequency range where the conductance decreases rapidly, the capacitance shows a sharp increase suggesting strong coupling between capacitance and conductivity (see Fig. 4.2). The maximum of the Cole-Cole plot is inversely proportional to the relaxation time. The semilogarithmic plot of this frequency ν_{max} versus the reciprocal of the absolute temperature, shown in the inset of Fig. 4.2, yields a straight line and shows that the relaxation process is thermally activated with an activation energy $\Delta E = 0.95$ eV. The only physical mechanism compatible with the observed behavior as a function of film thickness, electrode material, temperature and field (ac or dc) was that of dipole orientation involving the carbonyl functionalities which are incorporated into the polymer structure when it is exposed to the atmosphere via reaction of molecular oxygen with trapped free radical sites. The total concentration of these sites was estimated from the electrical data to be $\sim 4 \times 10^{20}$ cm^{-3} for the particular film used in good agreement with infrared data and ESR measurements on similar systems.

The observation of these dielectric relaxation processes arising from carbonyl features, has also been reported by Tibbit and co-workers[124)] in plasma polytetrafluoroethylene as well as other plama polymers. In measuring the dielectric loss tangents over a frequency range of 10^2–10^5 Hz at temperatures of −150 to 100 °C, they have demonstrated that the dielectric loss curves of plasma polymers derived from hydrocarbon and fluorocarbon monomers are very similar, but bear no resemblence to their conventionally polymerized counterparts.

5 Simultaneous Plasma Etching and Polymerization

The chemistry occurring in a plasma by an injected fluorocarbon is critically dependent upon the "effective" fluorine/carbon ratio in the system. We have seen that for fluorocarbons with a high F/C ratio, the predominant process is etching while for fluorocarbons with a low F/C ratio, oligomerization and polymerization are the dominant processes. In the past, these two areas of technology have been kept distinct and indeed the occurrence of polymerization in an etching system, for example, or vice-versa, has been considered a nuisance. In this section, therefore, we shall outline the potential utility of combining the two processes in the same system by reference to recent work by Kay and co-workers[65, 66] involving the incorporation of metals in plasma polymerized fluorocarbons by simultaneous etching and polymerization.

We have already inferred several methods by which a transition from etching to polymerization might be accomplished. These involved the effective lowering of the F/C ratio in the system by, for example, the addition of H_2 or C_2F_4 to a CF_4 plasma or even the effective removal of fluorine by conversion to stable volatile fluorides in the etching process. The transition in the reverse direction (from polymerization to etching) may also be achieved by, for example, for example, the addition of O_2 to a C_2F_4 plasma to effectively remove carbon as CO. In much of the work to be discussed below, C_3F_8 is employed in the system whose F/C stoichiometry allows convenient rates of both etching and polymerization to be maintained.

5.1 Metal Containing Fluorocarbon Polymers

The synthesis of the metal containing polymer films[66] involves a capacitively coupled diode reactor system in which one electrode is grounded (the anode). Radiofrequency power of 13.56 MHz is applied to the other electrode which at this frequency, attains an overall negative potential due to the greater mobility of the electrons than the ions in the plasma and is, therefore, termed the cathode. The positive ions, therefore, arrive at the cathode with increased kinetic energy and material is removed from its surface by competitive physical sputtering in which momentum transfer to the surface is involved, as well as, chemical plasma etching through the formation of volatile species which subsequently desorb and enter the gas phase. Silicon, germanium, molybdenum and tungsten are but a few examples of cathode materials which form volatile fluorides in a fluorocarbon plasma environment encompassing both non-metals and metals. Chemical plasma etching will be expected to dominate for these materials. Copper, on the other hand, is a typical example of a material which forms nonvolatile fluorides and thus material can only be removed from the cathode by a physical sputtering mechanism.

The effective removal of fluorine from the system by the etching process to form stable volatile fluorides tends to lower the effective fluorine/carbon ratio in the plasma and, thus, increase the concentration of unsaturated species and therefore the polymerization rate[45] (as discussed in a previous section) at the anode.

5.2 Structural Aspects of the Metal Containing Fluorocarbon Polymers

Extensive investigations of the structure of the films produced at the anode in systems as described above have been carried out by Dilks and Kay[65, 66]. In a series of experiments, the films synthesized by these methods employing germanium, molybdenum and copper cathodes, respectively, were compared. The ESCA data readily confirm that both molybdenum and copper may be incorporated into the fluoropolymer matrix formed at the anode although as previously noted, the method of cathode erosion must be different in the two cases (chemical plasma etching versus physical sputtering) (see Fig. 5.1). A consideration of the relative C_{1s}, F_{1s}, O_{1s} and metal core level (either Mo_{3d} or Cu_{2p}) intensities allows estimates of the empirical formula to be written for these materials as follows $[C_2F_{4.0}O_{0.6}Mo_{0.3}]_n$ and $[C_3F_{3.9}O_{0.3}Cu_{0.3}]_n$ representing ~19% molybdenum and 14% copper by weight, respectively, for the particular reactor employed[66]. The agreement of this surface analysis with data pertaining to the bulk composition obtained from X-ray fluorescence measurements[65,66], affirmed that the surface of these materials is representative of the bulk. Furthermore, transmission electron microscopy has showed that the metal entities in these materials are homogeneously dispersed throughout the film[66].

The overall structure of the films can be divided into two categories: the organic phase and the inorganic phase. Considering firstly the organic phase, comparison of the ESCA data (C_{1s} and F_{1s} levels) with data obtained previously for the plasma polymer derived from tetrafluoroethylene[114], shows a close similarity and it is, therefore, suggested that the two structures are probably derived from the same precursors to polymerization. Hence, it appears that the interaction of the perfluoropropane fragments with the cathode material in the etching process produces species in the gas phase with the general formula $(CF_2)_n$ which then polymerize by the mechanism suggested in the previous section. Mass spectrometric investigations of the plasma effluents provide strong evidence for this since the major new species on

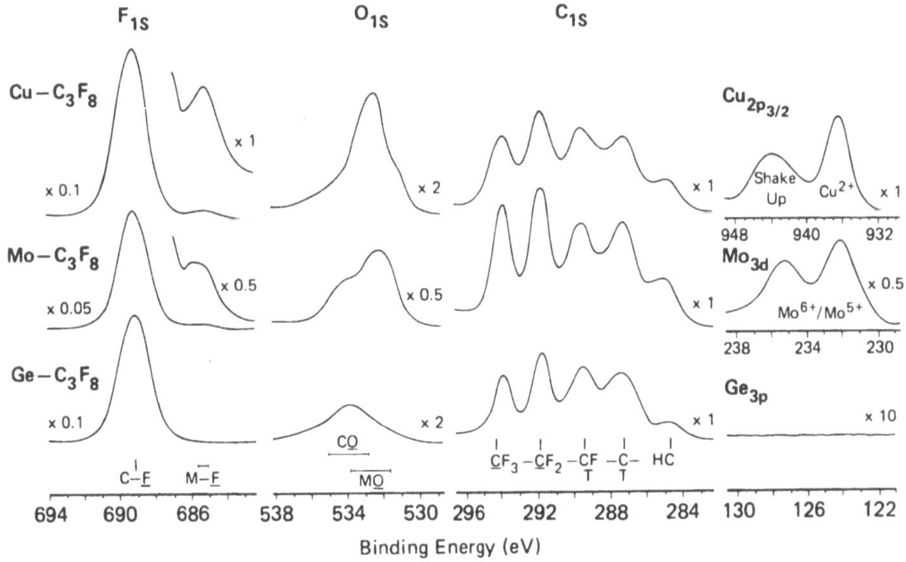

Fig. 5.1. ESCA spectra of the polymers formed in three experiments involving germanium, molybdenum and copper cathodes, respectively

exciting the plasma in perfluoropropane in systems employing germanium, molybdenum or copper cathodes were found to be $(CF_2)_n$ [66]. The relative concentration of these species in the gas phase is directly related to the polymer deposition rate which is in decreasing order Ge—C_3F_8 > Mo—C_3F_8 $\gg$ Cu—C_3F_8 (see Fig. 5.2), reflecting the abilities of the cathode etching processes to effectively reduce the F/C ratio in the plasma and thus produce polymerizable species. This is entirely reasonable since etch rate measurements[66] show that germanium is etched faster than molybdenum, in this system, and the physical sputtering mechanism for copper is expected to eject the copper from the cathode largely as atoms, due to the weak CuF bond strength[66], resulting in very little effective removal of fluorine from the system and, therefore, much less production of $(CF_2)_n$ polymerization precursors.

Despite the fact that germanium is etched at the faster rate for a given set of plasma conditions, it does not become incorporated into the polymer matrix[66] under the conditions of these experiments. This provides a convenient method to prepare films containing no cathode material but are essentially plasma polymerized polytetrafluoroethylene. The small oxygen signal observed in the ESCA spectrum of this

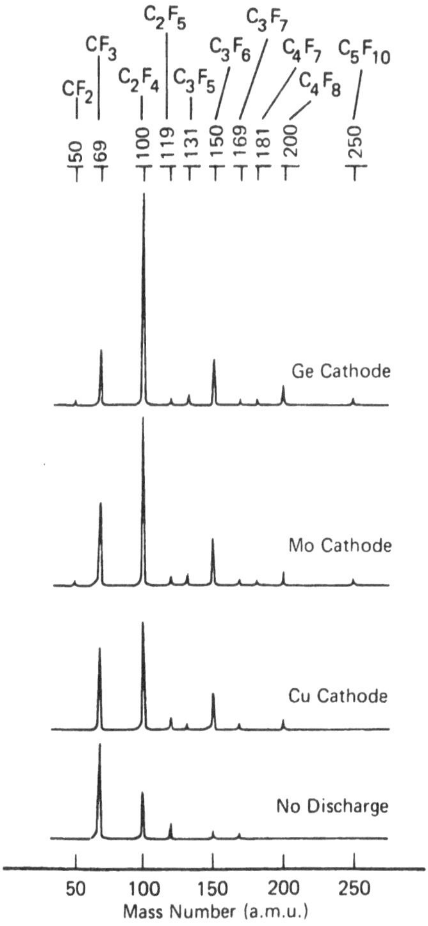

Fig. 5.2. Mass spectra of the plasma effluent neutral species, obtained with an electron impact ionization energy ~15 eV

material is then attributable to carbonyl functionalities produced by reaction of trapped free radical sites with molecular oxygen on exposure to the atmosphere. The concentration of these features is consistent with previous data pertaining to similar films ($\sim 10^{20}$ cm^{-3})[123].

On comparing this fluorocarbon film with those containing metal[66], a large increase in the total oxygen signal intensity is apparent for the metal containing films and the new oxygen features have a binding energy consistent with them being attached to the metal. This, along with the fact that ESCA reveals the metals to be present in their highest oxidation states, affirms that the metals are present as oxides. The oxygen again originates from the atmosphere, but two distinct mechanisms occur for molybdenum and copper. For the former, the metal is known from mass spectroscopy to arrive at the substrate as MoF_6, which subsequently becomes incorporated in the polymer matrix. Reaction of this entity with atmospheric moisture then forms molybdenum-oxygen bonds with the evolution of HF by a hydrolysis mechanism. For copper, on the other hand, the metal arrives at the substrate largely as atoms. Reaction of copper atoms in the polymer matrix with atmospheric oxygen permeating the film then produces CuO.

The general structure of this class of materials can, therefore, be summarized as a fine dispersion of metal oxide in a polymer matrix very similar to plasma polytetrafluoroethylene and in principle any metal should be able to be incorporated. Clearly, if the films are protected from the atmosphere, for metals which form involatile fluorides having a relatively weak metal-fluorine bond strength, it should be possible to produce films having metal atoms dispersed in the matrix. It is expected that these films will have many interesting chemical, optical, electrical and magnetic properties.

5.3 Control of Composition of Metal Containing Fluorocarbon Polymers

Experiments to determine the effects on the film structure and composition of varying the plasma parameters within moderate limits have also been reported for the system involving a molybdenum cathode and perfluoropropane as the injected monomer[66]. The important general features can be summarized as follows: (i) While the amount of metal incorporated into the films varies within narrow limits (18–26% molybdenum, by weight) with the flow rate of monomer, pressure and power, it is directly related to the ratio of the etch rate divided by the deposition rate. This narrow range of composition is not entirely surprising since the relative concentration of metal fluoride and polymerizable species must be closely related as they are both derived from the same etching mechanism which controls the F/C ratio in the gas phase. (ii) The structure of the metal containing entities is essentially unchanged over the range of plasma conditions studied. (iii) While the polymer structure varies very little as a function of the operating parameters, there is a distinct tendency for those formed at higher power loadings to be more highly crosslinked[66]. This is consistent with the fact that at higher power loadings, both the concentration and kinetic energy of species which are able to cause fluorine elimination in the forming film are increased.

Merely changing the operating parameters of the plasma, therefore, is not a satisfactory method to control the overall composition of the metal containing films[125], and it has been shown that more success is achieved by changing the relative rates of etching and deposition[125]. The latter may be accomplished by controlling the fluorine/carbon ratio in the plasma in such a way that increasing the ratio results in a greater metal incorporation due to an increase in etch rate with a concomitant decrease in deposition rate. Thus, substitution of C_2F_6, in place of C_3F_8, as the injected monomer produces films of higher metal content while the use of nC_4F_{10} produces films of lower metal content[125]. It is interesting to note that plasma effluent mass spectrometric studies have shown[125] that for C_2F_6 and nC_4F_{10}, the primary precursors to polymerization are also $(CF_2)_n$, although in the case of nC_4F_{10}, there is greater probability of these species being formed directly from the injected gas by electron impact processes[125]. The polymer structure for these films, therefore, is also related to that of plasma polytetrafluoroethylene. For experiments in which hydrogen or oxygen addition has been used to manipulate the effective F/C ratio, however, the polymer structure becomes modified by the inclusion of carbon-hydrogen or carbon-oxygen functionalities, respectively[125].

6 References

1. Plasma Diagnostics, Lochte-Holtgreven, W. (ed.)., New York, Elsevier, 1968
2. Plasma Diagnostic Techniques, Huddlestone, R. H., Leonard, S. S. L. (ed.)., New York, Academic Press, 1965
3. Podgornyi, I. M.: Topics in Plasma Diagnostics, New York, Plenum Press, 1971
4. Yamaguchi, S., Sawa, G., Ieda, M.: J. Appl. Phys. *48*, 2363 (1977)
5. Harrison, A. G. et al.: Can. J. Chem. *44*, 1967 (1966)
6. Schram, B. L. et al.: J. Chem. Phys. *44*, 49 (1966)
7. Kieffer, L. J.: Atomic Data *1*, 19 (1969)
8. Winters, H. F.: J. Chem. Phys. *44*, 1472 (1966)
9. Winters, H. F.: J. Chem. Phys. *63*, 3462 (1975)
10. Winters, H. F.: Chem. Phys. *36*, 353 (1979)
11. Blint, R. J., McMahon, T. B., Beauchamp, J. L.: J. Am. Chem. Soc. *96*, 1269 (1974)
12. Pabst, M. J. K., Tan, H. S., Franklin, J. L.: Int. J. Mass Spectrom. & Ion Phys. *20*, 191 (1976)
13. Krause, J. R., Lampe, F. W.: J. Phys. Chem. *81*, 281 (1977)
14. Heckel, E., Hanrahan, R. J.: J. Chem. Phys. *62*, 1027 (1975)
15. Marcotte, D. E., Hanrahan, R. J.: J. Fluorine Chem. *2*, 8 (1972/3)
16. Coburn, J. W., Chen, M.: J. Appl. Phys. *51*, 3134 (1980)
17. Schott, L.: In Ref. 1, p. 668
18. Swift, J. D., Schwar, M. J. R.: Electrical Probes for Plasma Diagnostics, New York, Elsevier, 1971
19. Chen, F. F.: In Ref. 2, p. 113
20. Chung, P. M., Talbot, L. Touryan, K. J.: Electric Probes in Stationary and Flowing Plasmas: Theory and Application, New York, Springer-Verlag, 1975
21. Clements, R. M.: J. Vac. Sci. Technol. *15*, 193 (1978)
22. Thomas, T. L., Battle, E. L.: J. Appl. Phys. *41*, 3428 (1970)
23. Szusczewicz, E. P., Holmes, J. C.: J. Appl. Phys. *46*, 5134 (1975)
24. Chang, J.-S.: J. Phys. D. *6*, 1674 (1973)
25. Barnes, B. T.: J. Appl. Phys. *33*, 3319 (1962)

26. Coburn, J. W., Winters, H. F.: J. Appl. Phys. *50*, 3189 (1979), and references in this article
27. Thornton, J. A.: J. Vac. Sci. Technol. *15*, 188 (1978)
28. Koenig, H. R., Maissel, L. I.: IBM J. Res. Develop. *14*, 168 (1970)
29. Coburn, J. W., Kay, E.: J. Appl. Phys. *43*, 4965 (1972).
30. Vossen, J. L.: J. Electrochem. Soc.: *126*, p. 319 (1979)
31. Coburn, J. W.: unpublished
32. Chapman, B. N., Minkiewicz, V. J.: J. Vac. Sci. Technol. *15*, 329 (1979)
33. Hasted, J. B.: Int. J. Mass Spectrom. & Ion Phys. *16*, 3 (1975)
34. Helm, H. et al.: J. Phys. B *7*, 170 (1974)
35. Smith, D., Plumb, I. C.: J. Phys. D *6*, 1431 (1973)
36. Franklin, J. L., Studniarz, S. A., Ghosh, P. K.: J. Appl. Phys. *39*, 2052 (1968)
37. Vasile, M. J., Smolinsky, G.: Int. J. Mass Spectrom. & Ion Phys. *12*, 133 (1973)
38. Rowe, B., Int. J. Mass Spectrom. & Ion Phys. *16*, 209 (1975)
39. Foner, S. N., Hudson, R. L.: J. Chem. Phys. *21*, 1374 (1953)
40. Fite, W. L.: Int. J. Mass Spectrom. & Ion Phys. *16*, 109 (1975)
41. Evans, H. E., Jennings, P. P.: Carbon *6*, 236 (1968)
42. Smolinsky, G., Vasile, M. J.: Int. J. Mass Spectrom. & Ion Phys. *12*, 147 (1973); *24*, 311 (1977)
43. Vasile, M. J., Smolinsky, G.: ibid. *13*, 381 (1974); *24*, 11 (1977)
44. Vasile, M. J., Smolinsky, G.: J. Phys. Chem. *81*, 2605 (1977)
45. Kay, E., Coburn, J. W., Kruppa, G.: Vide *183*, 89 (1976)
46. Coburn, J. W., Kay, E.: Proc. 7th Intern. Vac. Congr. & 3rd Intern. Conf. Solid Surfaces, Vienna *2*, 1257 (1977)
47. Stirling, A. J., Westwood, W. D.: J. Appl. Phys. *41*, 742 (1970)
48. Greene, J. E., Sequeda-Osorio, F.: J. Vac. Sci. Technol. *10*, 1144 (1973)
49. Bindley, T. F., Walker, S.: Trans. Faraday Soc. *58*, 217 (1962)
50. Stille, J. K., Sung, R. L., Van der Kooi, J.: J. Org. Chem. *30*, 3116 (1965)
51. Gilbert, R. Théorêt, A.: J. Phys. Chem. *80*, 1017 (1976)
52. Degenkolb, E. O. et al.: Appl. Spectrosc. *30*, 520 (1976)
53. Harshbarger, W. R. et al.: Appl. Spectrosc. *31*, 201 (1977)
54. Mogab, C. J.: J. Electrochem. Soc. *124*, 1262 (1977)
55. Mogab, C. J., Shankoff, T. A.: J. Electrochem. Soc. *124*, 1766 (1977)
56. Mogab, C. J., Adams, A. C., Flamm, D. L.: J. Appl. Phys. *49*, 3796 (1978)
57. Curtis, B. J., Brunner, H. J.: J. Electrochem. Soc. *125*, 829 (1978)
58. Oshima, M.: Japan J. Appl. Phys. *17*, 1157 (1978)
59. Chuang, T., Kay, E.: To be published
60. Eernisse, E. P.: J. Vac. Sci. Technol. *12*, 564 (1975)
61. Coburn, J. W., Winters, H. F., Chuang, T. J.: J. Appl. Phys. *48*, 3532 (1977)
62. Raby, B. A.: J. Vac. Sci. Technol. *15*, 205 (1978)
63. Horiike, Y., Shibagaki, M.: Extended Abstr., Proc. 151st Electrochem. Soc. *77:1*, 619 (1977)
64. Chuang, T. J.: Phys. Rev. Lett. *42*, 815 (1979)
65. Kay, E., Dilks, A., Hetzler, U.: Macromol. Sci.-Chem. *A12(9)*, 1393 (1978)
66. Kay, E., Dilks, A.: ACS Symp. Ser.: 108 "Plasma Polymerization" by Shen, M., Bell, A. T., Amer. Chem. Soc., 1979
67. Lehmann, H. W., Krausbauer, L., Widmer, R.: J. Vac. Sci. Technol. *14*, 281 (1977)
68. Melliar-Smith, C. M.: J. Vac. Sci. Technol. *13*, 1008 (1976)
69. Chemical Dry Etching System CDE-IV, Tokuda Seisakusho, Ltd., Kawasaki, Japan
70. Bersin, R. L., Singleton, M.: USP 3, 879, 597
71. Reinberg, A. R., in: Etching (Rand, M. J., Hughes, A. R. (ed.)), Princeton, N. J., Electrochem. Soc. Symp. Ser. 1976, p. 91
72. Hosokawa, N., Matsuzaki, N., Asamaki, T.: Japan J. Appl. Phys. Supp. 2, Part 1, 435 (1974)
73. Schwartz, G. C., Zielinski, L. B., Schopen, T., in: Etching (Rand, M. J., Hughes, A. R. (ed.)), Princeton, N. J., Electrochem. Soc. Symp. Ser. 1976, p. 122
74. Bondur, J. A.: J. Vac. Sci. Technol. *13*, 1023 (1976)
75. Jacob, A.: USP 3,867,216
76. Horiike, Y., Shibagaki, M.: Suppl. Japan J. Appl. Phys. *15*, 13 (1976)
77. Suzuki, K. et al.: ibid. *16*, 1979 (1977)

78. Pitts, J. N., Sandoval, J. J., Atkinson, R.: Chem. Phys. Lett. *29*, 31 (1974)
79. Atkinson, R. et al.: J. Geophys. Res. *81*, 5765 (1976)
80. Winters, H. F., Coburn, J. W., Kay, E.: J. Appl. Phys. *48*, 4973 (1977)
81. Heinecke, R. A. H.: Solid-State Electron. *18*, 1146 (1975)
82. Ephrath, L. M.: J. Electrochem. Soc.: 126, 1419 (1979)
83. Lehmann, H. W., Widmer, R.: Appl. Phys. Lett. *32*, 163 (1978); Appl. Phys. Lett. *33*, 367 (1978)
84. Lehmann, H. W., Widmer, R.: J. Vac. Sci. Technol. *15*, 319 (1978)
85. Coburn, J. W., Kay, E.: IBM J. Res. & Develop. *23*, 33 (1979)
86. Bersin, R. L., Reichelderfer, R. F.: Solid State Technol. *20(4)*, 78 (1977)
87. Winters, H. F.: J. Appl. Phys. *49*, 5165 (1978)
88. Matsuo, S., Takehara, Y.: Japan J. Appl. Phys. *16*, 175 (1977)
89. Matsuo, S.: ibid. *17*, 235 (1978)
90. Holland, L., Ojha, S. M.: Vacuum *26*, 53, 233 (1976)
91. Holland, L.: J. Vac. Sci. Technol. *14*, 5 (1977)
92. Mauer, J. L. et al.: ibid. *15*, 1734 (1978)
93. Hanak, J. J., Pellicane, J. P.: ibid. *13*, 406 (1976)
94. Cuomo, J. J., et al.: ibid. *15*, 281 (1978)
95. Schaible, P. M., Metzger, W. C., Anderson, J. P.: J. Vac. Sci. Technol. *15*, 334 (1978)
96. Millard, M. M., in: Techniques and Applications of Plasma Chemistry, (Hollahan, J. H., Bell, A. T. (Ed.)), John Wiley & Sons, New York, 1974
97. Shen, M. (Ed.): Plasma Chemistry of Polymers. Marcel Dekker, Inc., New York, 1976, and J. Macromol. Sci.-Chem. *A10* (1976)
98. Clark, D. T., Dilks, A., Shuttleworth, D., in: Polymer Surfaces (Clark, D. T., Feast, W. J. (Ed.)), John-Wiley & Sons, London, in press
99. Yasuda, H., in: Thin Film Processes (Vossen, J. L., Kern, W. (Ed.)), Academic Press, New York, 1978
100. Tien, P. K., Smolinsky, G., Martin, R. J.: Appl. Opt. *11*, 637 (1972)
101. Wydeven, T., Kubadhi, R.: Appl. Opt. *15*, 132 (1976)
102. For example, Yasuda, H.: Appl. Pol. Symp. *22*, 241 (1973)
103. Stancell, A. F., Spencer, A. T.: J. Appl. Pol. Sci. *16*, 1505 (1972)
104. Dubois, J. C., Gazard, M., Zann, A.: J. Appl. Phys. *47*, 1270 (1976)
105. Sprokel, G. J.: Mol. Cryst. Liq. Cryst. *42*, 223 (1977)
106. Bui, H., Carchano, H., Sanchez, D.: Thin Solid Films *13*, 207 (1972)
107. Ozawa, P. J.: IEEE Trans. Parts. Mater. Package PMP-5 *112* (1969)
108. Yasuda, H., Hsu, T. S.: J. Pol. Sci., Pol. Chem. Edn. *15*, 2411 (1977)
109. Kay, E., Coburn, J. W., Kruppa, G.: VIDE *183*, 89 (1976)
110. Biederman, M., Ojha, S. M., Holland, L.: Thin Solid Films *41*, 329 (1977)
111. Kobayashi, H., Bell, A. T., Shen, M.: Macromolecules *7*, 277 (1974)
112. Dacey, J. R., Littler, J. G. F.: Can. J. Chem. *47*, 3871 (1969)
113. Maserangelo, S. V. R.: J. Amer. Chem. Soc. *84*, 1122 (1962)
114. Rice, D. W., O'Kane, D. F.: J. Electrochem. Soc. *123*, 1308 (1976)
115. Clark, D. T., Dilks, A.: J. Pol. Sci., Pol. Chem. Edn. *16*, 911 (1978), and references therein
116. Seeger, M. et al.: J. Polym. Sci., Polymer Chem. Edn. *15*, 1403 (1977)
117. Fyfe, C. A. et al.: Macromolecules *12*, 757 (1979)
118. O'Kane, D. F., Rice, D. W.: J. Macromol. Sci.-Chem. *A10*, 567 (1976)
119. Millard, M. M., Windle, J. J., Pavlath, A. E.: J. Appl. Pol. Sci. *17*, 2501 (1973)
120. Millard, M. M., Pavlath, A. E.: J. Macromol. Sci.-Chem. *A10*, 579 (1976)
121. Millard, M. M., in: Characterization of Metal & Polymer Surfaces, Vol. 2. (Lee, L. M. (ed.)), Academic Press, New York, 1977
122. Diaz, A. et al.: J. Amer. Chem. Soc. *99*, 6780 (1977)
123. Hetzler, U., Kay, E.: J. Appl. Phys. *49*, 5617 (1978)
124. Tibbit, J. M., Bell, A. T., Shen, M.: J. Macromol. Sci.-Chem. *A10*, 519 (1976)
125. Kay, E., Dilks, A., Seybold, D.: J. Appl. Phys. 1980 in press.

The Mechanism and Kinetics of Plasma Polymerization

Alexis T. Bell

Department of Chemical Engineering, University of California, Berkeley, CA 94720

Table of Contents

1 Introduction

Rather unusual high molecular weight products can be formed by passage of an organic vapor through a plasma produced in a low pressure electrical discharge. This process, known as plasma polymerization, is applicable to a very wide range of starting materials including those not normally considered to be monomers for conventional polymerization. In contrast to conventional polymerization, plasma polymerization produces an irregular network-like material which is amorphous and highly crosslinked. Moreover, the stoichiometric composition of the plasma-deposited polymer can differ significantly from that of the starting monomer. The possibility of preparing such unique structures has prompted efforts to find applications for plasma-deposited films. Examples of some of the more promising areas of application are the production of thin-film capacitors, reverse osmosis membranes, optical waveguides, antireflection coatings, protective coatings, and biocompatible coatings.

A fairly substantial body of literature is now available relating to the kinetics of plasma polymerization and the physical and chemical characteristics of the deposited material. Since these topics have been summarized in several reviews[1-7], they will not be discussed any further here. Instead, the present review will focus on two objectives. The first is to summarize the available evidence concerning the mechanisms of plasma polymerization. In this context the roles of electrons and ions in initiating polymerization and the nature of the active species contributing to chain propagation will be examined. Specific reference will be made to the polymerization of hydrocarbon monomers in order to limit the scope of this portion of the discussion. The second objective will be to review the theoretical models which have been proposed to describe the kinetics of plasma polymerization. A brief discussion of the physical characteristics of low pressure discharges will also be presented to clarify the influence of discharge operating conditions on these characteristics.

2 Reactor Configurations

Plasma polymerization is usually carried out in a low pressure glow discharge sustained by either a dc or an ac electric field. Examples of the reactors used for this purpose are shown in Fig. 1. The simplest configuration involves a pair of circular parallel plate electrodes mounted inside a glass bell jar. The lower electrode usually serves as the substrate holder and is sometimes heated or cooled. Monomer is introduced through a feed tube and unconsumed monomer and gaseous products are withdrawn through a port in the base plate.

A significant disadvantage of the bell jar configuration is that the monomer flow is not constrained to pass totally through the plasma formed between the electrodes. To overcome this problem, the electrodes can be mounted in a rectangular flow channel. By adding channel sections before and after the plasma zone, it is further possible to establish a well-developed flow profile for the gas as it enters the plasma zone. This reactor design is particularly well suited for studies of film deposition kinetics.

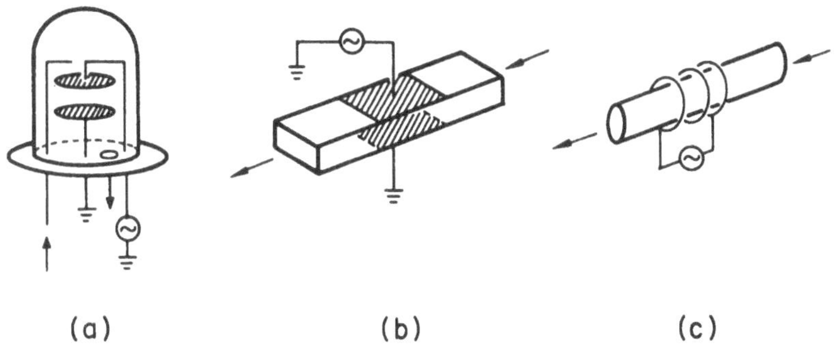

Fig. 1. Reactors used for plasma polymerization: **a** bell jar; **b** rectangular flow channel; **c** electrode-less

An electrodeless reactor configuration is also shown in Fig. 1. In this instance the plasma is coupled to the power source through a coil wound on the outside of the reactor. In order to achieve efficient coupling between the plasma and the power supply, the supply frequency must be in the rf frequency range. The principal advantage of the electrodeless design is that it eliminates the contamination which might arise from electrodes placed within the plasma. If the discharge is operated so that the plasma extends beyond the region defined by the coil, then it is possible to excite an inert gas in the intense plasma zone within the coil and to introduce monomer in the glowing plasma which spreads downstream. This method of monomer introduction has some advantage where it is desired to avoid extensive degradation of the monomer.

3 Physical Characteristics of the Plasma

The physical characteristics of a discharge and the manner in which it is sustained can have a profound effect on the kinetics of plasma polymerization[8]. Therefore, we shall review these topics here, with specific emphasis on the characteristics of plasmas sustained between parallel plate electrodes. This constraint is imposed because virtually all efforts to theoretically model the kinetics of plasma polymerization have been directed towards plasmas of this type. Readers interested in broader and more detailed discussions of plasma characteristics can find such in references[9-11].

If a sufficiently high dc potential is imposed across a pair of electrodes, the gas filling the space between the electrodes will break down and a self-sustained discharge will be established. The potential, field, space charge, and current density distributions characteristic for such a discharge are illustrated in Fig. 2. The high net positive space charge present in the cathode dark space causes a sudden increase in the potential between the cathode and the leading edge of the negative glow. This part of the potential is referred to as the cathode fall and is typically 100 to 400 V in magnitude. Because of the high potential gradient in the cathode dark space, electrons leaving the cathode are accelerated to high energies before

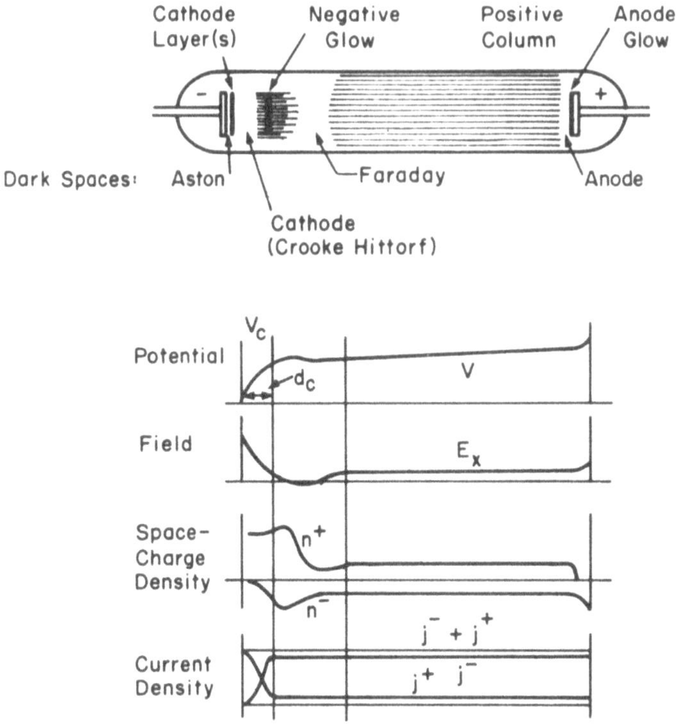

Fig. 2. Distribution of voltage, field, space charge, and current density in a dc glow discharge

entering the negative glow. The portion of the discharge between the negative glow and the end of the positive column is characterized by a small potential increase and nearly equivalent concentrations of ions and electrons. Because the potential gradient in these regions is considerably smaller than in the cathode dark space, the electron energies are also smaller. Finally, in the space between the positive column and the anode a third potential increase is observed. This component is referred to as the anode fall and is typically 10 to 20 V in magnitude. It arises due to the excess negative space charge present in the anode dark space.

Since electrons and ions are swept out of the discharge by the imposed field, they must be replaced continuously. New electrons are created at the cathode by the impingement of ions accelerated through the cathode fall and by photons created in the negative glow. The majority of positive ions collected by the cathode are replaced by electron-impact ionization occurring in the negative glow.

When a low frequency (< 100 Hz) field is used to create the discharge, each electrode alternately acts as cathode and anode. On each half cycle a dc-type discharge is established once the breakdown potential is surpassed. As long as the applied voltage remains above the maintenance potential, the discharge is sustained, but when the voltage falls below the maintenance potential, the discharge is extinguished, and the space charge is swept out of the volume. For sufficiently low frequencies all of the space charge is eliminated before the potential rises on the next half cycle to a point where a new discharge is initiated.

As the frequency of the applied field is increased, a point is reached at which the time taken by a positive ion to move between the electrodes becomes equal to one half the period of the field. The corresponding frequency is defined by[10]

$$f_{c_i} = \frac{\langle v \rangle_{d_i}}{2L}, \tag{1}$$

where $\langle v \rangle_{d_i}$ is the average ion drift velocity and L is the interelectrode distance. For frequencies above f_{c_i}, ions created near a momentary anode cannot transit to the cathode before the field is reversed. This leads to a partial retention of the positive space charge from one half cycle to the next and facilitates reinitiation of the discharge. For systems which do not depend upon secondary electron generation through ion bombardment, increasing the frequency above f_{c_i} has no significant effect on the starting or maintenance potential of the discharge[10].

Raising the frequency of the applied field above f_{c_i} leads to the observation of a second critical frequency f_{c_e}, corresponding to the condition that an electron makes the transit between electrones in one half cycle of the field. The relationship defining f_{c_e} is[10]

$$f_{c_e} = \frac{\langle v \rangle_{d_e}}{2L} \tag{2}$$

where $\langle v \rangle_{d_e}$ is the average electron drift velocity. For frequencies above f_{c_e} both the positive and negative changes are retained from one half cycle to the next, leading to a significant reduction in the loss or charged particles from the system. As a result, the voltages required to both initiate and maintain the discharge decrease abruptly[9-11].

The spatial distributions of electrons and ions in a discharge maintained at frequencies above f_{c_e} are governed by the processes controlling the formation and loss of these species. Since the ac field no longer causes significant displacement of either electrons or positive ions, the loss of charged species from the discharge is controlled by ambipolar diffusion and homogeneous recombination[9,10]. Replacement of charged species occurs by electron-impact ionization of neutral gas molecules. Throughout the bulk of the discharge volume the concentrations of electrons and ions are essentially equal. The concentration profiles have a maximum at the midpoint between the electrodes and fall off monotonically towards each electrode. Within a Debye length or so of the electrode surface a space charge sheath is set up across which a potential of several volts may exist. The presence of the sheath accelerates the transport of ions but decelerates the transport of electrons. Thus, even for requencies above f_{c_e} there is a flux of ions and electrons to the electrode surface. However, these fluxes are orders of magnitude smaller than those experienced when the transport is mobility controlled.

For frequencies between f_{c_i} and f_{c_e}, the distribution of space charge and potential within the discharge is governed by the manner in which the applied field affects the motions of ions and electrons. Because of their low mobility the positive ions are unable to move significant distances during each half cycle of the field. This results in an essentially stationary spatial distribution of ions which has a maximum value at the center of the discharge. In contrast to the behaviour of the ions, the electrons are swept through the interelectrode gap by the field on each half cycle, causing the electron distribution to be a function of both position and time. This situation can give rise to a pulsating negative sheath voltage, especially when the electrodes are covered by a dielectric film, as will occur in a discharge used for plasma polymerization. Due to the difference in mobility between electrons and ions, the current-voltage characteristics of the discharge will resemble those of a leaky rectifier, as shown in Fig. 3[12,13]. Upon initial application of an rf voltage across the dielectric film, a high electron current flows to the surface of the film (Fig. 3a). On the second half the cycle, only a relatively small ion current can flow. Since no charge can be transferred through the dielectric film, the voltage on the film surface must self-bias negatively until the net current (averaged over each cycle) is zero. This results in the pulsating negative potential shown in Fig. 3b. The average dc value of this potential

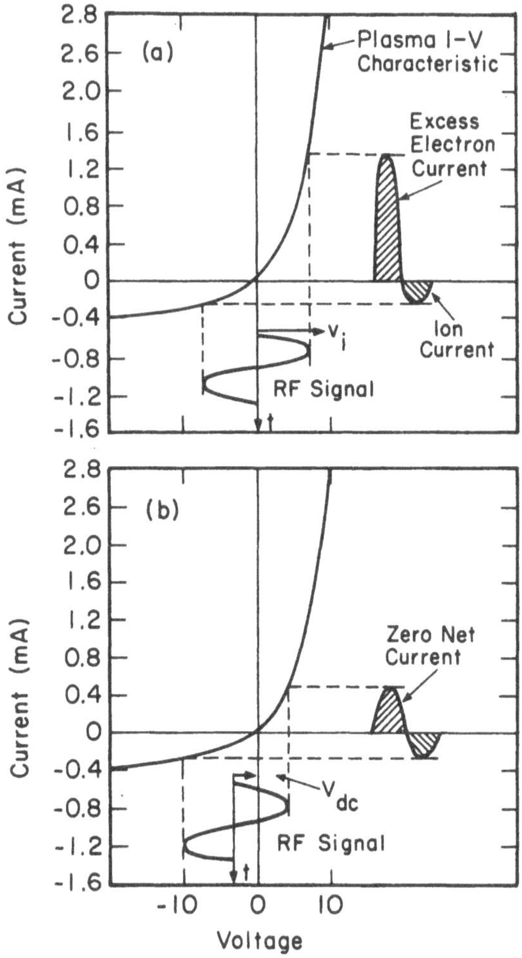

Fig. 3. Formation of a pulsating negative sheath on a capacitively coupled surface in an rf glow discharge [after Butler and Kino[12]]: **a)** initial application of rf voltage; **b)** steady-state application of rf voltage

is nearly equal to the peak voltage applied and is significantly larger than the sheath potential observed in discharges operated at frequencies above f_{c_e} [14].

It should be noted, however, that while the frequency of the power source used to sustain the discharge does affect the flow of charge to the electrode surfaces and the potential across the plasma-electrode sheath, the characteristics of the positive column are little affected by frequency. Within the positive column the average electron energy lies between 1 and 10 eV and is determined almost exclusively by E_e/p, the ratio of the effective electric field strength and the gas pressure[9–11,15]. The effective electric field strength is in turn related to the amplitude of the electric field strength, E_0, by the relationship

$$E_e = \frac{E_0}{\sqrt{2}} \left(\frac{\nu^2}{\nu^2 + \omega^2} \right)^{1/2}, \tag{3}$$

where $\omega = f/2\pi$ and ν is the frequency of elastic collisions between electrons and gas molecules. For a positive column in which electron loss occurs primarily by ambipolar diffusion, the magnitude of E_e/p is dictated by $p\Lambda$, where Λ is a characteristic diffusion length[9–11,15].

4 Elementary Processes Involved in Plasma Polymerization

It is particularly difficult to define a mechanism of plasma polymerization because of the very large number of elementary processes which must be considered. As a consequence, most authors have sought to interpret the kinetics of polymer deposition in terms of a very simple reaction scheme encompassing only a small number of steps. This has resulted in significant differences of opinion regarding the nature of the species which propagate chain growth and the processes involved in initiating polymerization.

While it has been suggested that plasma polymerization can be propagated by cationic species[16–19], an increasing body of evidence[8,20–23] indicates that free radicals are the dominant species responsible for chain growth. Several factors argue in favor of this conclusion. Firstly, it is recognized that at pressures of a torr and above the free radical concentrations in a nonequilibrium plasma are usually 10^3 to 10^5 higher than the ion concentrations. As a consequence it can be shown[8] that free radical-molecule reactions occur at rates 10 to 10^2 faster than ion-molecule reactions. Secondly, it has been observed[20] that the addition of small amounts of haloforms to hydrocarbon monomers causes a significant acceleration of the monomer polymerization rate and a decrease in the H/C ratio of the polymer. These effects are most easily interpreted by assuming that free halogen atoms released in the plasma enhance the free radical population via hydrogen abstraction. A third reason for preferring a free radical mechanism is the observation of high radical concentrations at the surfaces of polymers exposed to a plasma[21,22]. These surface free radicals are capable of reacting with unsaturated monomers[23].

In the balance of this section we shall review the types of elementary reactions believed to be responsible for plasma polymerization. For purposes of clarity these reactions have been listed in Table 1 and grouped into five generic types-initiation, adsorption, propagation, termination, and reinitiation. The factors controlling the rate of each type of reaction will also be discussed.

Table 1. Elementary Processes Occurring During Plasma Polymerization

Initiation	Termination
1. $e + M_g \rightarrow 2R_g + e$ 2. $e + M_g \rightarrow M_g' + 2H \text{ (or } H_2) + e$ 3. $M_s \xrightarrow{(I^+, e^- h\nu)} 2R_s$	9. $R_g \text{ (or H)} + R_g' \rightarrow P_g$ 10. $R_g + R_s' \rightarrow P_s$ 11. $R_s + R_s' \rightarrow P_s$
Adsorption	**Reinitiation**
4. $M_g + S \rightleftarrows M_s$ 5. $R_g + S \rightleftarrows R$	12. $e + P_g \rightarrow R_g + R_g' + e$ 13. $H + P_g \rightarrow R_g + H_2$ 14. $P_s \xrightarrow{(I^+, e^-, h\nu)} R_s + R_s'$
Propagation	
6. $R_g \text{ (or H)} + M_g \rightarrow R_g'$ 7. $R_s + M_g \rightarrow R_s'$ 8. $R_s + M_s \rightarrow R_s'$	

As indicated in Table 1, free radicals can be formed in the gas phase by the collision of energetic free electrons with monomer molecules. And in a related process, hydrogen atoms are produced together with a hydrogen depleted monomer molecule. The ease with which this process occurs appears to increase with increasing saturation of the monomer[6]. Once formed, the hydrogen atoms can produce further free radicals by either hydrogen abstraction or addition to an olefin.

The rates of both electron-monomer collision processes can be described by

$$r_{i_g} = k_{i_g} [e] [M] \tag{4}$$

where [e] and [M] are the concentrations of electrons and monomer, respectively. The rate coefficient k_{i_g} depends upon the cross section for the dissociative process of interest, the average electron energy or temperature, and the shape of the electron energy distribution function[8,15].

Primary free radicals can also be formed on the surface of the deposited polymer through the dissociation of adsorbed monomer caused by the energy released upon impact of electrons and ions or the absorption of photons emanating from the plasma. The rates at which free radicals are formed by these processes can be expressed as

$$r^{i}_{i_s} = \sigma^{i}_{i_s}\Gamma_i [M_s] , \quad (5)$$

$$r^{e}_{i_s} = \sigma^{e}_{i_s}\Gamma_e [M_s] , \quad (6)$$

$$r^{p}_{i_s} = \sigma^{p}_{i_s}\Gamma_p [M_s] , \quad (7)$$

where Γ is the flux of particles or photons reaching the polymer surface and $[M_s]$ is the surface concentration of adsorbed monomer. The cross section σ will be different for each process and strongly dependent on the energy of the incoming particle or photon. A more detailed discussion of reactions similar to those described here is given by Winters[24)].

The adsorption of both monomer and gas phase free radicals is expected to occur at the surface of the polymer film in contact with the plasma. If equilibrium is established between a species in the gas phase and the equivalent adsorbed species, then the surface coverage by the jth species can be expressed as

$$[X_{sj}] = \frac{K_j[X_{gj}]}{1 + \sum_j K_j[X_{gj}]} , \quad (8)$$

assuming a Langmuir form for the isotherm. The quantities $[X_{s_j}]$ and $[X_{g_j}]$ represent the surface and gas phase concentrations of the jth species and K_j the equilibrium constant. The magnitude of K_j is associated with the entropy and heat of adsorption of the adsorbing species, and will normally increase with decreasing surface temperature.

In the event that equilibrium is not achieved, the rates of adsorption and desorption will be given by

$$r_a = k_a[X_{g_j}] , \quad (9)$$

and

$$r_d = k_d [X_{s_j}] , \quad (10)$$

respectively. The adsorption rate coefficient, k_a, in Eqn. 9 is proportional to the product of the random thermal velocity of the jth species and its sticking coefficient. On the other hand, the desorption rate coefficient, k_d, is equivalent to the reciprocal of the average residence time of the jth species on the polymer surface prior to desorption.

The propagation of chain growth can occur both in the gas phase and on the surface of the depositing polymer film. In the gas phase propagation may involve the addition of either a hydrogen atom or a free radical to an alkene or alkyne. It should be noted that the hydrocarbons which propagate chain growth may be either the original monomer fed to the plasma or unsaturated products formed by electron-molecule collisions. The rate at which gas phase propagation occurs can be expressed as

$$r_{p_g} = k_{p_g} [H] [M_g] \tag{11}$$

or

$$r_{p_g} = k_{p_g} [R_g] [M_g] . \tag{12}$$

At the polymer surface, propagation can occur by the reaction of surface free radicals with either gas phase or adsorbed monomer. The rates for these processes can be expressed as

$$r_{p_{sg}} = k_{p_{sg}} [R_s] [M_g] , \tag{13}$$

or

$$r_{p_{ss}} = k_{p_{ss}} [R_s] [M_s] . \tag{14}$$

The rate coefficients $k_{p_{sg}}$ and $k_{p_{ss}}$ can not be evaluated readily and hence are usually treated as empirical parameters.

The termination of chain growth can also occur both in the gas phase and at the polymer surface. In the gas phase, free radicals are lost by reaction with both hydrogen atoms and other free radicals. The kinetics of these processes are given by

$$r_{t_g} = k_{t_g} [H] [R_g] , \tag{15}$$

or

$$r_{t_g} = k_{t_g} [R_g'] [R_g] . \tag{16}$$

At the polymer surface radicals are lost by reactions involving gaseous atomic hydrogen, gas phase free radicals, and adsorbed free radicals. The rate of surface termination can be expressed as

$$r_{t_{sg}} = k_{t_{sg}} [H] [R_s] , \tag{17}$$

$$r_{t_{sg}} = k_{t_{sg}} [R_g] [R_s] , \tag{18}$$

$$r_{t_{ss}} = k_{t_{ss}} [R_s'] [R_s] . \tag{19}$$

The rate coefficients appearing in Eqns. 17 through 19 should not be strongly temperature dependent since radical-atom and radical-radical recombination are most often either unactivated or weakly activated processes[25)]. In the case of the recombination of two surface free radicals, the rate is likely to be limited by the mobility of the polymer chains attached to the radicals. For very short chains, as are commonly produced in plasma polymerization[26)], only those radicals which are nearest or next nearest neighbors are likely to react. If one of the radicals is secondary and the other primary, recombination will result in a polymer branch point. If both radicals are secondary, then a cross-link will be formed.

The chain fragments formed by the recombination of free radicals can be reconverted into radicals by a variety of reinitiation processes, some of which are listed in Table 1. Such reactions can occur in the gas phase via electron collision and on the polymer surface by impact of charged particles or photon absorption. Reinitiation may also be induced in both the gas phase and on the polymer surface by hydrogen transfer reactions. These last processes are similar to the chain transfer processes which occur during homogeneous polymerization. Expressions for the rates of reinitiation are given by Eqns. 20 through 23.

$$r_{ri_g} = k_{ri_g} [e] [P_g] , \tag{20}$$

$$r_{Ht_g} = k_{Ht_g} [H] [P_g] , \tag{21}$$

$$r_{ri_s} = \sigma_{ri_s} \Gamma_i [P_s] , \tag{22}$$

$$r_{Ht_s} = k_{Ht_s} [H] [P_s] . \tag{23}$$

5 Models of Plasma Polymerization Kinetics

In the present section we shall review the attempts which have been made to model quantitatively the kinetics of plasma polymerization. The assumptions underlying each model will be discussed as well as the extent to which the predictions of the theoretical models fit the experimental data. At tne end of this section it will be shown why the initial assumptions made in developing kinetic models depend on the conditions used to sustain the plasma.

One of the earliest attempts to interpret the kinetics of plasma polymerization was presented by Williams and Hayes[27]. Working with a plasma sustained at 10 kHz between parallel-plate electrodes, they observed that the rate of polymer deposition on the electrodes rose to a saturation level as either the gas pressure or the discharge current was increased. This evidence suggested that active species were formed on the electrode surface rather than in the gas phase. It was proposed that polymerization is initiated by the formation of active species through the impact of ions with the polymer film covering the electrodes. Propagation was envisioned to occur via the reaction of surface active species with adsorbed monomer vapor. The role of adsorbed monomer was supported by the observation that at a fixed pressure the rate of polymerization of different monomers correlated with the relative vapor pressure of the monomer and that the rate of polymerization of a given monomer increased with decreasing electrode temperature.

A number of the mechanistic features proposed by Williams and Hayes were incorporated into a theoretical model developed by Denaro et al.[28] to explain the kinetics of styrene polymerization in a 2 MHz discharge. Initiation was proposed to proceed through the collision of electrons with the polymer film covering the electrodes. Ion collisions were excluded because at 2 MHz and a pressure of around one torr ions would not be able to follow the applied electric field and, hence, would move only by diffusion. The energy transferred to the polymer film by inelastic electron collisions was assumed to produce free radicals either by direct excitation or by ionization followed by excitation into a dissociative state. Both processes are shown below.

1. $P_s \xrightarrow{e^-} P_s^* \rightarrow 2R_s$,

2. $P_s \xrightarrow{e^-} P_s^+ + e^- \rightarrow P_s^{**} \rightarrow 2R_s$.

The free radicals produced by steps 1 and 2 were postulated to react via the following scheme:

3. $R_{s_n} + M_s \rightarrow R_{s_{n+1}}$,

4. $R_{s_n} + R_{s_m} \rightarrow P_{s_{n+m}}$ (or $P_{s_n} + P_{s_m}$) ,

5. $R_{s_n} \rightarrow \dot{R}_{s_n}$ (trapped) .

From this scheme they deduced the rate expression

$$r_P = r_R \frac{k_2[R_s][M_s]}{k_3[R_s][M_s] + k_4[R_s]^2 + k_5[R_s]} \tag{24}$$

where r_R is the rate of formation of radicals.

Rearrangement of Eqn. 24 and introduction of the relationship $[M_s] = Kp$, where p is the monomer vapor pressure, leads to

$$\frac{1}{r_P} = \frac{1}{p}\left\{\left(\frac{k_4}{Kk_3}[R_s] + \frac{k_5}{Kk_3}\right)\Big/ r_R\right\} + \frac{1}{r_R}. \tag{25}$$

A plot of the reciprocal of the experimentally observed deposition rate versus the reciprocal gas pressure was found to be linear for a given discharge power. From the intercepts of such plots it was deduced that r_R depended upon the power W according to the relationship $r = k'W^{1.7}$, suggesting that the production of radicals is based on a more complex mechanism than single stage bombardment by electrons.

By applying the stationary state treatment to the formation of radicals, a value for $[R_s]$ could be obtained. Substitution of this expression into Eqn. 24 resulted in the final form of the deposition rate.

$$r_P = \frac{pk'W^{1.7}}{p + A + (A^2 + Bk'W^{1.7})^{1/2}} , \tag{26}$$

where

$$A = k_3/(2Kk_1) , \qquad B = k_2/(Kk_1)^2 ,$$
$$= 0.277 , \qquad = 0.011 .$$

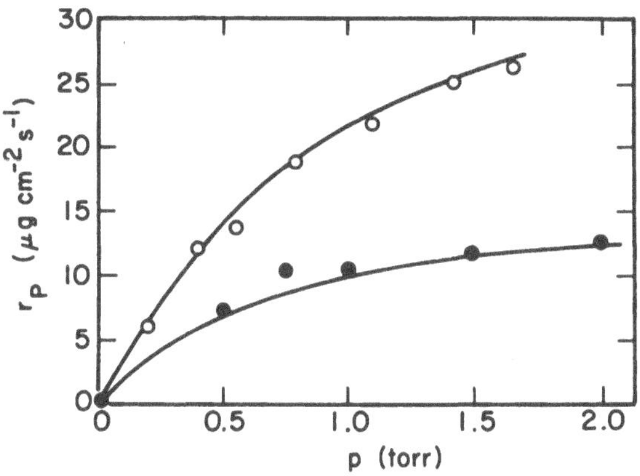

Fig. 4. Comparison of calculated and experimental results. Experimental: • — 1.5 W/cm^2; ○ — 2.7 W/cm^2. Calculated: — [after Denaro et al.[28]]

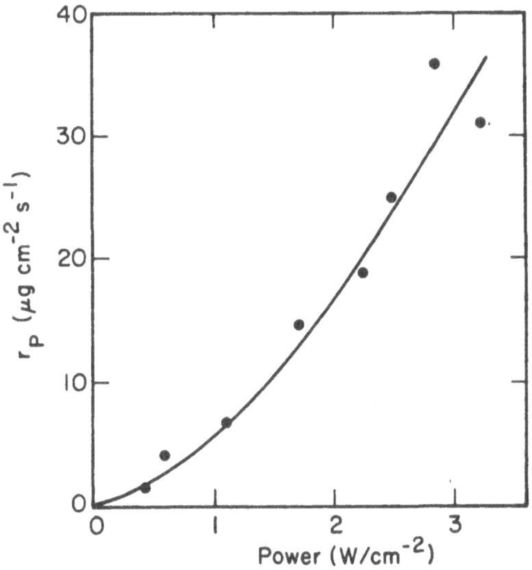

Fig. 5. Comparison of calculated and experimental results for 1.4 torr. Experimental: •. Calculated: — [after Denaro et al.[28]]

Figures 4 and 5 show that the rate of styrene polymerization predicted by Eqn. 26 is in good agreement with that observed experimentally.

In a subsequent pair of studies Denaro et al.[29,30] showed that a simplified version of Eqn. 26 given by

$$r_p = k'W^n \frac{p}{p + 2A}, \tag{27}$$

could be used to correlate the rates of polymerization of styrene, α — methyl styrene, allylbenzene, allyl alcohol, and crotyl alcohol. Values of the parameters appearing in Eqn. 27 are given in Table 2.

A model very similar to that developed by Denaro et al.[28] was also proposed by Carchano[31]. In this case surface free radicals were taken to be formed via

Table 2. Rate Parameters for Plasma Polymerization of Different Monomers

Monomer	Parameters[a]		
	k'	n	A
Styrene	8.5	1.7	0.29
α-Methyl-styrene	12.3	1.0	0.20
Allyl benzene	12.0	1.0	0.55
Allyl alcohol	5.8	0.57	1.02
Crotyl alcohol	1.4	1.0	0.26

[a] Parameter values assuming p in torr, W in W/cm^2, and r_P in $\mu g/cm^2$ sec.

activation of adsorbed monomer under the impact of ions or radicals. Subsequent propagation and termination were envisioned to occur as described by reactions 3 and 4, in Denaro et al.'s mechanism[28] but trapping of free radicals was not considered. This scheme led to a rate expression of the form

$$r_P = r_R \frac{1}{1 + (2k_3/k_1k_2)\,[R_s]\,(1/p)} \,. \tag{28}$$

Verification of Eqn. 28 was suggested by the linearity of a plot of r_P^{-1} versus $(p/p_0)^{-1}$, the reciprocal of the reduced pressure. It should be noted that p/p_0 was varied by changing the electrode temperature while maintaining the gas pressure constant. No attempt was made to propose a functional form for r_R.

The role of ions in initiating polymerization was further elaborated by Poll et al.[32]. For frequencies below f_{c_i} they proposed that the impact of ions with the growing polymer film produced free radicals via reactions 1 and 2. New polymer would then form by the reaction of surface free radicals with adsorbed monomer. It was also proposed that the concentration of surface free radicals is proportional to the current density striking the growing polymer surface and, hence, that the rate of polymerization is given by

$$r_R = \sigma^i_{i_s} \Gamma_i [M_s] \,. \tag{29}$$

The steady state value of $[M_s]$ is given by

$$[M_s] = \gamma\Gamma_M/(\tau^{-1} + \sigma^i_{i_s}\Gamma_i + \gamma\Gamma_M/[M_s]_0) \,, \tag{30}$$

where Γ_M is the flux of monomer molecules striking the polymer surface, γ is the monomer sticking coefficient, τ is the average residence time for an adsorbed monomer molecule prior to desorption, and $[M_s]_0$ is the maximum concentration of adsorbed monomer. Substituting Eqn. 30 into Eqn. 29 and assuming that $\gamma\Gamma_M\tau \gg [M_s]_0$ leads to the following expression for r_P:

$$r_P = \frac{\gamma\Gamma_M}{1 + \gamma\Gamma_M/(\Gamma_i\,[M_s]_0)} \,. \tag{31}$$

The form of Eqn. 31 leads to the expectation that r_P will approach an assymptotic value as either Γ_i or Γ_M are increased, consistent with experimental observations[27,28,32].

To account for the effects of gas phase composition on the rate of polymer deposition, Poll et al.[32] proposed the following reaction network:

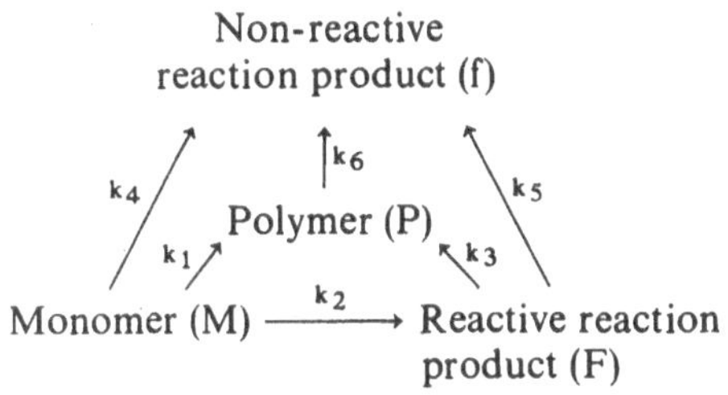

This scheme suggests that monomer can be converted into reactive and nonreactive products through processes occurring in the plasma (reaction pathways 2 and 4) as well as entering into polymer formation (reaction pathway 1). The reactive products may further contribute to polymer deposition (reaction pathway 3) or be converted to non-reactive products (reaction pathway 5). The degradation of the polymer to form non-reactive products (reaction pathway 6) is also considered.

Species balances written for each of the components lead to the following set of differential equations:

$$\frac{dP}{dt} = \frac{A_1M}{1 + B_1M} + \frac{A_2F}{1 + B_2F} - k_6P_0\,, \tag{32}$$

$$\frac{dM}{dt} = -\frac{A_1M}{1 + B_1M} - k_2M - k_4M - k_7M + Q\,, \tag{33}$$

$$\frac{dF}{dt} = -\frac{A_2F}{1 + B_2F} + k_2M - k_5F - k_7F\,, \tag{34}$$

$$\frac{df}{dt} = k_6P_0 + k_4M + k_5F - k_7f\,. \tag{35}$$

The constants A_j and B_j ($j = 1, 2$) are given by

$$A_j = \gamma v_j\,, \tag{36}$$

$$B_j = A_j/(\sigma^i_{i_s}\Gamma_i[M_s]_0)\,, \tag{37}$$

where v_j is the average thermal velocity of M or F. The rate coefficient k_7 takes account of the evacuation of monomer and reactive products, and Q the influx of new monomer. The constant P_0 describes the amount of polymer present prior to initiation of polymerization.

Solutions of Eqns. 32 through 35 on an analog computer are shown in Fig. 6 for the case of polymerization in a closed system. The plots of $r_p(t)$ and $p(t)$ are seen to be in qualitative agreement with the experimental observations shown in Fig. 7. The model of polymerization kinetics was also found to provide curves of r_p versus discharge current density and monomer flow rate which were consistent qualitatively with the experimentally observed results.

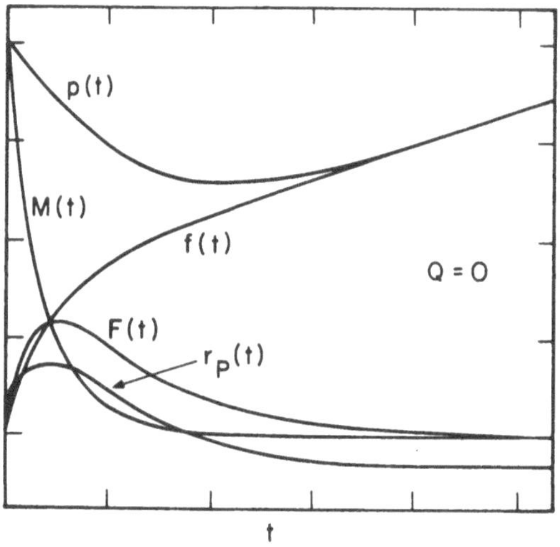

Fig. 6. Analog computer solutions for p(t), M(t), f(t), F(t), and $r_p(t)$ [after Poll et al.[32]]

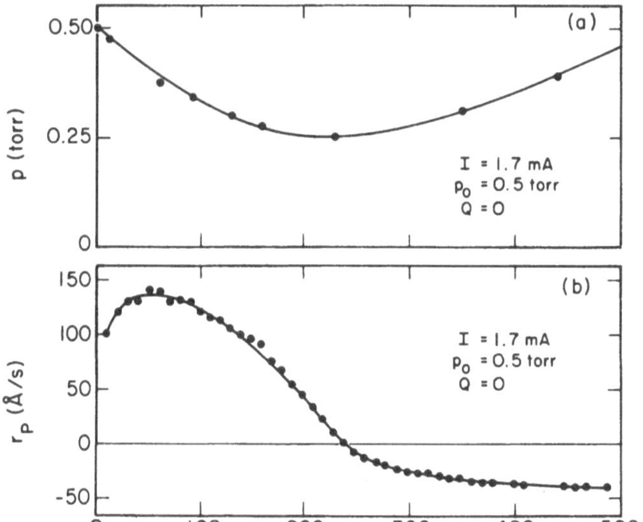

Fig. 7. Dependence of total pressure (**a**) and film growth rate (**b**) on glow discharge duration in a closed system (Q = 0; P_0 = 0.5 torr; I = 1.7 mA) (monomer C_2F_4). [after Poll et al.[32]]

The deposition of polymeric films by plasma polymerization of styrene in a 800 kHz discharge was investigated by Lam et al.[33]. It was proposed that the observed deposition kinetics could be explained by a scheme in which the initiation of monomers by electron impact is followed by propagation and termination, as in conventional polymerization. This scheme is summarized by the following three reactions:

Initiation: $M + e \rightarrow 2R_1 + e$

Propagation: $R_n + M \rightarrow R_{n+1}$

Termination: $R_m + R_n \rightarrow P_{m+n}$ or $P_m + P_n$

Four possible deposition models were then considered, each of which is outlined below.

Model 1. Gas-phase initiation, propagation, and termination.
Model 2. Gas-phase initiation and propagation; surface termination.
Model 3. Gas-phase initiation; surface propagation and termination.
Model 4. Surface initiation, propagation, and termination.

For each of the four cases an expression for the rate of polymer deposition was derived in terms of the gas pressure, p, the plasma current, I, the measure of the average electron energy, V/p, and the ratio of the monomer pressure to its saturation vapor pressure, x. The functional dependencies of r_P on these variables are given in Table 3. The constants a, c, and n, are taken to be parameters whose values are adjusted to obtain a fit between the measured and predicted values of r_P.

Table 3. Functional Dependece of Polymer Deposition Rate

Model	Initiation	Propagation	Termination	r_P
1	Gas	Gas	Gas	$ap^{3/2}I^{1/2}(V/p)^n$
2	Gas	Gas	Surface	$ap^3I(V/p)^n$
3	Gas	Surface	Surface	$ap^{1/2}I^{1/2}(V/p)^n \times \{cx/(1-x)[1+(c-1)x]\}$
4	Surface	Surface	Surface	$aI^{1/2}(V/p)^n \times \{cx/(1-x)[1+(c-1)x]\}$

To evaluate each rate expression appearing in Table 3, the values of the a, c, and n were varied to obtain a minimum in the square of the relative error between the theoretically predicted and experimentally measured deposition rates. The results of this nonlinear regression analysis are presented in Table 4. It can be seen that only for Models 3 and 4 is $|\delta_{max}|$ smaller than the errors in the measurements, $\pm 9.5\%$[33].

The suitability of Models 3 and 4 was further examined by comparing the rates predicted by each model as a function of the substrate temperature, T_s. The measured ratios of the rate determined at T_s to the rate measured at 20 °C are listed in Table 5 together with the ratios projected by Models 3 and 4. It is evident that Model 3 consistently predicts the ratio of the rates more accurately than

Table 4. Regression Analysis Results

Rate Expression Model	Parameters[a]			Percentage	Difference
	$a \times 10^3$	n	c	$\lvert\delta_{max}\rvert$	δ_{rms}
1	5.0	0.65	—	32	20
2	0.80	0.70	—	92	58
3	5.5	0.70	19	7.8	4.7
4	15	0.58	5.5	8.1	4.9

[a] Parameter values assuming p in torr, I in mA, V in V(rms), and r_p in μg/cm² min.

Model 4. As a result Lam et al. concluded that Model 3 best describes the plasma polymerization kinetics of styrene.

The role of gas phase initiation processes was further explored by Tibbitt et al.[34]. These authors proposed that the polymerization of unsaturated hydrocarbons in a 13.56 MHz plasma is initiated by free radicals formed in the gas by electron-monomer collisions, the initiation reactions listed in Table 6. Moreover, it was assumed that the formation of free radicals on the polymer surface due to the impact of charged particles could be neglected. This assumption is supported by the fact that at 13.56 MHz and pressures near one torr the discharge frequency is significantly greater than either f_{c_i} or f_{c_e} and that as a result the fluxes of charged particles to the electrode surfaces are quite small.

The rest of the mechanism proposed by Tibbitt et al.[34] is also shown in Table 6. Two assumptions were introduced to simplify the mechanism. The first is that the extent of gas phase termination is very small and consequently, that essentially all of the radicals formed in the gas phase are adsorbed on the polymer surface. The second assumption is that the concentrations of adsorbed monomer and free radical are proportional to the gas phase concentrations of these species. These relationships are expressed by

$$[M_s] = K_M[M_g]\,, \tag{38}$$

$$[R_s] = K_R[R_g]\,. \tag{39}$$

Table 5. Results of Adsorption Test on Model 3 and 4

T_s (°C)	p_0 (Torr)	p (Torr)	$r_p\lvert T_s/r_p\rvert_{20}$	$[M_s]_{T_s}/[M_s]_{20}$	
				Model 3	Model 4
16	4.5	0.45	1.13	1.12	1.22
16	4.5	0.65	1.08	1.11	1.21
24	7.0	0.45	0.97	0.91	0.85

Table 6. Reaction Mechanism for Plasma-Polymerization of Unsaturated Hydrocarbons

Initiation	Propagation (Heterogeneous)
1. $e + M_g \rightarrow M'_g + H_2 + e$	10. $R_{s_n} + \begin{Bmatrix} M_g \\ M'_g \end{Bmatrix} \rightarrow R_{s_{n+1}}$
2. $e + M_g \rightarrow M'_g + 2H + e$	
3. $e + M_g \rightarrow 2R_g + e$	11. $R_{s_n} + \begin{Bmatrix} M_s \\ M'_s \end{Bmatrix} \rightarrow R_{s_{n+1}}$
4. $e + H_2 \rightarrow 2H + e$	
Propagation (Homogeneous)	**Termination**
5. $H + \begin{Bmatrix} M_g \\ M'_g \end{Bmatrix} \rightarrow R_{g_1}$	12. $H_s + H \rightarrow H_2$
	13. $R_{g_m} + H \rightarrow P_{g_m}$
6. $R_{g_n} + \begin{Bmatrix} M_g \\ M'_g \end{Bmatrix} \rightarrow R_{g_{n+1}}$	14. $R_{g_m} + R_{g_n} \rightarrow P_{g_{m+n}}$
Adsorption	15. $R_{g_m} + R_{s_n} \rightarrow P_{s_{m+n}}$
	16. $R_{s_m} + R_n \rightarrow P_{s_{m+n}}$
7. $S + \begin{Bmatrix} M_g \\ M'_g \end{Bmatrix} \rightarrow \begin{Bmatrix} M_s \\ M'_s \end{Bmatrix}$	**Reinitiation**
	17. $e + P_{g_{m+n}} \rightarrow R_{g_m} + R_{g_n}$
8. $S + H \rightarrow H_s$	
9. $S + R_{g_n} \rightarrow R_{s_n}$	18. $P_s \xrightarrow{e,\, h\nu,\, I^+} R_{s_m} + R_{s_n}$
	19. $H + P_{g_n} \rightarrow R_{g_n} + H_2$
	20. $H + P_{s_n} \rightarrow R_{s_n} + H_2$

Using the assumptions just discussed, the rate of polymer deposition can be expressed as the rate of monomer consumption and is given by

$$r_P = \frac{L}{2} k_6[M_g][R_g] + k_{10}[M_g][R_s] + k_{11}[M_s][R_s],$$
$$= \left(\frac{L}{2} k_6 + k_{10}K_R + k_{11}K_MK_R\right)[M_g][R_g]. \quad (40)$$

To evaluate Eqn. 40 the concentrations of monomer and free radicals must first be known. These concentrations were determined by solving one-dimensional species conservation equations. The form of these equations is given by

$$\frac{d[X_j]}{dV} = \frac{[N_g]}{F}\left(r_j - \frac{[X_j]}{[N_g]} r_0\right), \quad (41)$$

where $[X_j]$ is the concentration of species j, $[N_g]$ is the total gas concentration, F is the total molar flow rate, r_j is the net rate of formation of species j by chemical reactions, and V is the volume of the plasma. The quantity r_0 is given by

$$r_0 = \sum_j r_j. \quad (42)$$

Since the total molar flow rate changes with axial position additional equation

$$\frac{dF}{dV} = r_0 \tag{43}$$

must be solved together with Eqn. 40.

To reduce the number of unknown rate coefficients appearing in Eqns. 41 and 43, it was assumed that $k_i[e] = k_1[e] = k_2[e] = k_4[e]$, $k_{p_g} = k_5 = k_6$, $k_a = k_8[S] = k_g[S]$, and $k_{p_s} = k_{10}K_R$, and that the rates of all other processes listed in Table 6 are negligibly small. Equations 41 and 43 were then solved for a chosen set of rate parameters and the local values of $[R_g]$ and $[M_g]$ were used to predict the local rate of polymer deposition. Integration of the local rate over the length of the plasma gave an overall rate of deposition which could be compared with experimentally measured values. If the agreement between experiment and theory was poor, the vector of rate coefficients was varied to obtain a good fit.

Figure 8 illustrates a comparison between measured and computed rates of butadiene polymerization in an rf plasma sustained at 13.56 MHz. A perfect fit is achieved by adjusting the rate coefficients appearing in the model to the following values:

$$k_i[e] = 2.1 \times 10^{-2}\ s^{-1},$$

$$k_{p_g} = 3.5 \times 10^{-18}\ cm^3/s,$$

$$k_a = 7.5 \times 10^{-3}\ cm/s,$$

$$k_{p_s} = 5.5 \times 10^{-18}\ cm^4/s.$$

For the conditions shown in Fig. 8, it was estimated that the electron density was $8 \times 10^9\ cm^{-3}$. This means that $k_i = 2.5 \times 10^{-12}\ cm^3/s$, a value in good agreement with measured rate coefficients for dissociation of small molecules by electron impact[15]. The gas phase propagation rate coefficient k_{p_g} was also found to be in very good agreement with values determined for conventional butadiene polymerization. The agreement of the adjusted parameter values with those measured independently lends further support to the validity of the proposed model of plasma polymerization.

Tibitt et al.[34] were also able to obtain an analytical expression for the rate of polymerization by assuming that hydrogen atoms are equivalent to free radicals. The final form of r_P is given by

$$r_P = \frac{2ca}{(a-b)} [M_g]_0^2 \frac{1 - e^{-(a+b)\tau}}{e^{(a+b)\tau}}, \tag{44}$$

where $a = k_i[e]$; $b = (2/L)\,k_a$; $c = (L/2)\,(k_{p_g} + k_{p_s})$; and $\tau = V/Q$. The quality of the fit of Eqn. 44 to experimental data is shown in Fig. 9 and values of the rate coefficient required are given in Table 7.

It is apparent from the preceding discussion that there are differences of opinion concerning the mechanism by which plasma polymerization is initiated: some

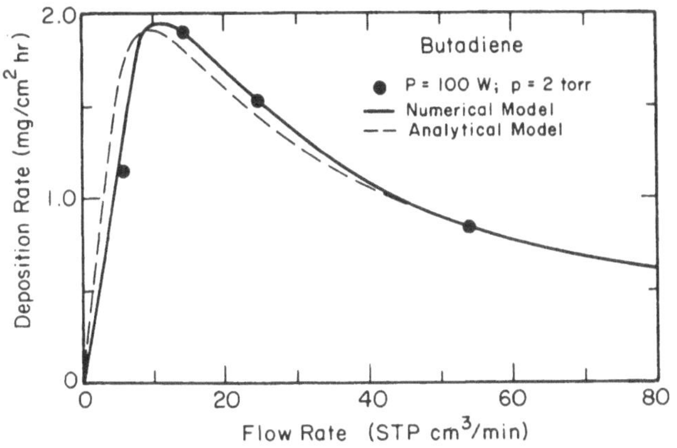

Fig. 8. Calculated and experimental rates of polymer deposition as a function of monomer flow rate [after Tibbitt et al.[34]]

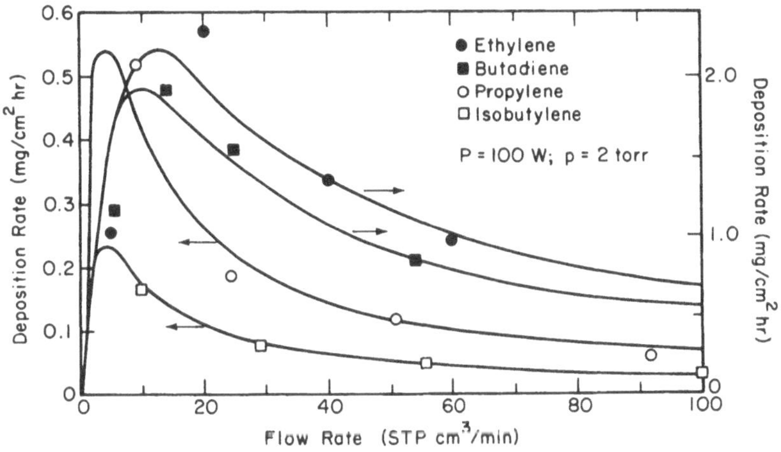

Fig. 9. Calculated and experimental polymer deposition rates for olefins as a function of monomer flow rate [after Tibbitt et al.[34]]

authors[21-32] suggesting that initiation occurs by ion and electron bombardment of the polymer surface and others[33,34] proposing that initiation occurs via electron-monomer collisions in the gas phase. The extent to which each of these mechanisms prevails is suggested by the recent work of Morita et al.[35]. These authors investigated the rate of ethane polymerization over the frequency range of 50 Hz to 13.56 MHz. As seen in Fig. 10, the rate of polymer deposition at a constant discharge power was found to be strongly dependent on the frequency, with significantly higher rates being observed at frequencies below 6 MHz.

Table 7. Fitted Rate Coefficients for the Plasma Polymerization of Unsaturated Hydrocarbon Monomers

Monomer	p (Torr)	W (W)	$k_i[e]$ (s^{-1})	$(2/L)\ k_a$ (s^{-1})	$(L/2)(k_{p_g} + k_{ps})$ (cm^4/s)
Acetylene	0.5	50	1.75	3.2×10^{-3}	2.45×10^{-16}
Ethylene	0.5	50	0.18	3.2×10^{-3}	2.40×10^{-17}
Isobutylene	2.0	100	0.03	3.2×10^{-3}	4.60×10^{-19}
Propylene	2.0	100	0.03	3.2×10^{-3}	1.45×10^{-18}
Butadiene	2.0	100	0.08	3.2×10^{-3}	3.65×10^{-18}
Ethylene	2.0	100	0.10	3.2×10^{-3}	7.70×10^{-18}
Ethylene	0.7	100	0.22	3.2×10^{-3}	2.90×10^{-17}
Ethylene	2.0	100	0.10	3.2×10^{-3}	7.70×10^{-18}
Ethylene	0.5	50	0.18	3.2×10^{-3}	2.40×10^{-17}

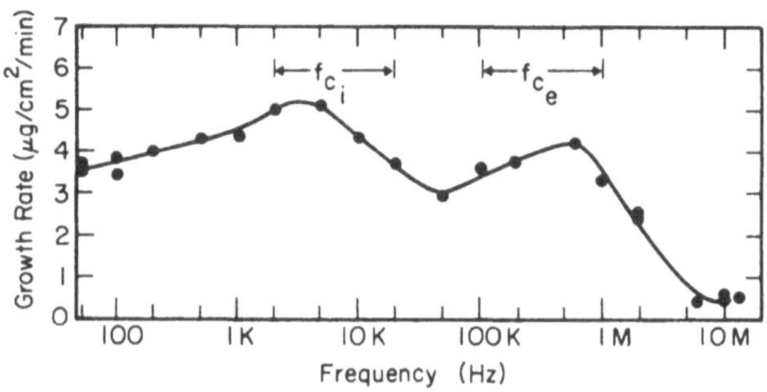

Fig. 10. Effect of discharge frequency on the rate of ethane plasma polymerization [after Morita et al.[35]]

The pattern observed in Fig. 10 can be explained in terms of the following mechanism for the polymerization of ethane.

1. $e + M \rightarrow 2R_g + e$,

2. $P_s \xrightarrow{e,\ I,\ h\nu} 2R_s$,

3. $R_g + P_s \rightarrow R_s$,

4. $R_g + R_s \rightarrow P_s$.

Surface free radicals are assumed to be produced by the dissociation of bonds at the polymer surface as well as by the adsorption of radicals from the gas phase. New polymer is formed by the recombination of surface and gas phase free radicals. Within the context of this model, the rate of polymerization should depend upon the rates at which surface and gas phase free radicals are produced. The influence of the discharge frequency on the rates of these processes can now be examined.

For frequencies below 5 kHz the discharge has the characteristics of a pulsating dc discharge. On each half cycle the electrode acting as the cathode is subjected to a flux of energetic ions accelerated, through the cathode fall. The impact of these ions on the surface of the polymer film covering the momentary cathode will cause a disruption of the polymer film and the formation of surface free radicals[36]. A similar phenomenon will occur at the opposite electrode, or anode, due to the impact of electrons. The formation of radicals is expected to be far more efficient at the cathode due to the high energy with which ions strike the cathode.

Gas phase free radicals will be formed throughout the discharge. However, due to the significantly higher field strength in the region of the cathode dark space, the rate of radical production will be maximized there. This is advantageous since it shortens the distance for radical diffusion from the gas phase to the cathode surface.

Because the discharge is initiated and extinguished on each half cycle, the rate of radical production is not constant but, rather, occurs in pulses. The time averaged concentration of gas phase radicals will depend upon the frequency of the pulses and the kinetics of radical consumption. In studies of pulsed photopolymerization it has been observed that the time averaged radical concentration will increase from a low level plateau to a high level plateau as the pulse frequency is increased[37]. This behavior causes the rate of polymerization to exhibit a similar response to increasing pulse frequency. A very similar trend has been observed in recent studies of ethylene and ethane polymerization using a pulsed rf discharge[38]. The rate of polymerization was found to increase by about 50% as the pulse frequency was increased from 500 sec^{-1} to 5000 sec^{-1}. These observations suggest that the increase in polymer deposition rate between 50 Hz and 5 KHz, observed in Fig. 10, might be ascribed to an increasing time averaged concentration of gas phase free radicals.

As the frequency of the applied field becomes greater than f_{c_i}, ion bombardment of the electrodes will be reduced. However, complete suppression of the ion flux will not occur until frequencies considerably greater then f_{c_i} are attained, due to the fact that ions located near the electrodes will still be able to reach the electrode surfaces in times small compared to $(2f_{c_i})^{-1}$. Over the same frequency range the effect of frequency on the gas phase concentration of free radicals is not expected to be large, since the time between individual discharge periods is too small for the radical population to collapse. As a result, it is anticipated that the rate of ion impingement will control the rate of polymer deposition and that a reduction in the impingement rate will cause a reduction in the rate of polymer deposition. The data shown in Fig. 10 support this contention since a sudden reduction in the polymer deposition rate is observed for frequencies greater than f_{c_i}.

The increase in polymer deposition rate which occurs between 50 and 600 KHz is the most difficult feature of Fig. 10 to explain. It is conceivable that this increase is associated with an increase in the energy of the electrons striking the electrode acting as the anode. Such an increase could occur if the magnitude of the anode fall were to increase. Thus, here again it is suggested that the effect of frequency is associated with the rate at which surface free radicals are produced.

Finally, it is proposed that the very low polymer deposition rates observed at frequencies near 10 MHz can be ascribed to low rates of formation of both surface and gas free radicals. At frequencies in this range, charge transport to the electrode surfaces is governed totally by ambipolar diffusion. As noted earlier, this transport mechanism leads to substantially lower current densities than can be achieved by field driven transport. Furthermore, the energies of the species arriving at the electrodes are much lower than those characteristics of frequencies below f_{c_e}. Because of these effects it is expected that the rate of surface free radical generation by the impact of charged species will be low.

As it was discussed earlier, the production of gas phase free radicals occurs via electron-impact dissociation at a rate which is proportional to the rate coefficient for dissociation and the electron density. The dissociation rate coefficient is strongly dependent on the electron temperature or its measure, E/p, which in turn is related to the discharge voltage and the gas pressure. At the same time the electron density is proportional to the discharge current. As a consequence, in order to determine the influence of frequency on the rate of gas-phase free radical formation, it is necessary to examine first the influence of frequency on the discharge voltage and current. Figure 11 illustrates these characteristics and shows that for frequencies below f_{c_e}, the voltage rises sharply and the current falls, in order to maintain a constant discharge power. These patterns suggest that below f_{c_e}, the values of E/p and [e] in the positive column are nearly constant and that above f_{c_e} the value of E/p increases and the value of [e] decreases with increasing discharge frequency. As a consequence of these trends it is anticipated that below f_{c_e} the rate of gas-phase free radical production will be independent of frequency. Above f_{c_e} the rising voltage is expected to increase the magnitude of the dissociation rate coefficient but the falling current decreases the electron density. The net effect on the dissociation rate is hard to judge. However, based upon the dissociation kinetics of other small molecules[15], it is expected that the rate will either stay the same or decline with an increase in frequency above f_{c_e}.

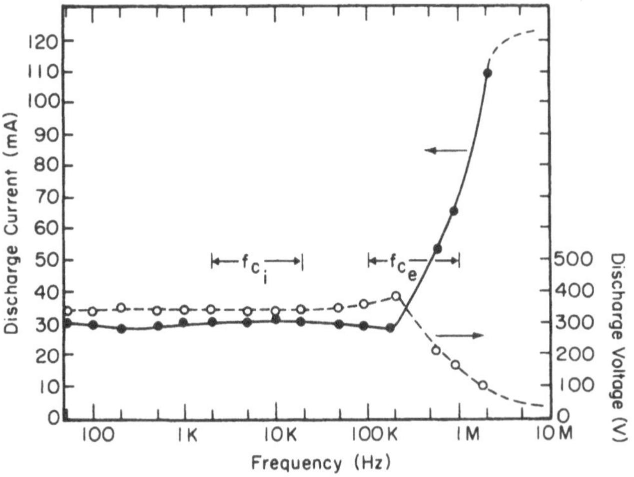

Fig. 11. Effect of discharge frequency on the voltage and current for an ethane plasma [after Morita et al.[35]]

In summary, then, the results presented in Fig. 10 suggest that the rate of polymer deposition is significantly influenced by the rate at which surface free radicals are formed through the impact of charged species with the growing polymer film. This process is complimented by the adsorption of gas phase free radicals. However, since the latter process appears to be slow, a reduction in the rate of surface radical production by the impact of charged species greatly reduces the rate of polymer deposition. These considerations lead to the conclusion that the primary mechanism of initiation is strongly dependent on the frequency used to sustain the plasma, a factor which must be recognized when considering the modelling of plasma polymerization kinetics.

6 Conclusions

The discussion presented in this review has shown that plasma polymerization is a complex process involving a large number of both homogeneous and heterogeneous reactions. While the details of the overall polymer deposition mechanism are not yet understood fully, certain features have been elucidated. Thus, the available experimental evidence suggests that free radicals are the primary species propagating chain growth, both in the gas phase and on the surface of the deposited polymer. These species are formed in the gas phase through the collision of free electrons with monomer molecules and on the surface of the growing polymer film through the impact of ions and electrons. Surface free radicals are also produced through the adsorption of a part of the radicals formed in the gas phase. Subsequent growth of the polymer film occurs by the reaction of surface-free radicals with either gas-phase free radicals or unsaturated monomer.

The number of attempts to model the kinetics of plasma-polymerization has been limited thus far. Nevertheless, these efforts have been useful in demonstrating the role of different processes in initiating polymerization and the manner in which the physical characteristics of the plasma affect the polymerization rate. It is anticipated that future modeling efforts will provide more detailed descriptions of the polymer deposition kinetics and thereby aid the development of a better understanding of the interactions between the physical characteristics of a plasma and the chemistry associated with polymer formation.

7 References

1. Kolotyrkin, V. M., Gilman, A. B., Tsapuk, A. K.: Russ. Chem. Rev. *36*, 579 (1967)
2. Mearns, A. M.: Thin Solid Films *3*, 201 (1969)
3. Millard, M., in: Techniques and Applications of Plasma Chemistry (Hollahan, J. R., Bell, A. T. ed.). New York, Wiley-Interscience, 1974
4. Plasma Chemistry of Polymers (Shen, M. ed.), New York, Marcel Dekker, 1976
5. Havens, M. R., Biolsi, M. R., Mayhan, K. G.: J. Vac. Sci. Technol. *13*, 575 (1976)

6. Yasuda, H., in: Thin Film Processes (Vossen, J. L., Kern, W. ed.), New York, Academic Press, 1978
7. Plasma Polymerization, ACS Symp. Ser. 108, (Shen, M., Bell, A. T. ed.), Washington, D. C., Amer. Chem. Soc., 1979
8. Bell, A. T., in: Plasma Chemistry of Polymers (Shen, M. ed.), New York, Marcel Dekker, 1976
9. Brown, S. C.: Basic Data of Plasma Physics, Cambridge, Mass., M.I.T. Press, 1959
10. Francis, G.: Ionization Phenomena in Gases, London, Butterworths, 1960
11. Nasser, E.: Fundamentals of Gaseous Ionization and Plasma Electronics, New York, Wiley — Interscience, 1971
12. Butler, H. S., Kino, G. S.: Phys. Fluids *6*, 1346 (1963)
13. Anderson, G. S., Mayer, W. N., Wehner, G. K.: J. Appl. Phys. *37*, 574 (1966)
14. Tsui, R. T. C.: Phys. Rev. *168*, 107 (1968)
15. Bell, A. T., in: Techniques and Applications of Plasma Chemistry, (Hollahan, J. R., Bell, A. T. ed.), New York Wiley — Interscience, 1974
16. Haller, I., White, P.: J. Phys. Chem. *67*, 1784 (1963)
17. Thompson, L. F., Mayhan, K. G.: J. Appl. Polym. Sci. *16*, 2291 (1972)
28. Thompson, L. F., Mayhan, K. G.: J. Appl. Polym. Sci. *16*, 2317 (1972)
19. Smolinsky, G., Vasile, M. J.: J. Macromol. Sci., Chem. *10*, 473 (1976)
20. Kobayashi, H., Shen, M., Bell, A. T.: J. Macromol. Sci., Chem. *8*, 1345 (1974)
21. Morosoff, N., Crist, B., Bumgarner, M., Hsu, T., Yasuda, H.: J. Macromol. Sci., Chem. *10*, 451 (1976)
22. Morita, S., Migutani, T., Ieda, M.: Japn. J. Appl. Phys. *10*, 1275 (1971)
23. Bradley, A., Fales, D.: Chem. Technol. 232 (April 1971)
24. Winters, H. F., in: Plasma Chemistry III — Topics in Current Chemistry (Venugopalan, M., Vepřek, S. ed.). Berlin, Springer Verlag, 1960
25. Kondrat'ev, V. N.: Chemical Kinetics of Gas Reactions, Reading, Mass., Addison-Wesley, 1964
26. Tibbitt, J. M., Shen, M., Bell, A. T.: J. Macromol. Sci., Chem. *10*, 1623 (1976)
27. Williams,T., Hayes, M. W.: Nature *209*, 769 (1966)
28. Denaro, A. R., Owens, P. A., Crawshaw, A.: Europ. Polymer J. *4*, 93 (1968)
29. Denaro, A. R., Owens, P. A., Crawshaw, A: ibid. *5*, 471 (1969)
30. Denaro, A. R., Owens, P. A., Crawshaw, A.: ibid. J. *6*, 487 (1970)
31. Carchano, H.: J. Chem. Phys. *61*, 3634 (1974)
32. Poll, H. U., Artz, M., Wickleder, K. H.: Europ. Polymer J. *12* 505 (1976)
33. Lam, D. K., Baddour, R. F., Stancell, A. F., in: Plasma Chemistry of Polymers (Shen, M. ed.), New York, Marcel Dekker, 1976
34. Tibbitt, J. M., Jensen, R., Bell, A. T., Shen, M.: Macromol. *10*, 647 (q977)
35. Morita, S., Bell, A. T., Shen, M.: J. Polym. Sci. Polym. Chem. Ed. *17*, 2775 (1979)
36. Hudis, M., in: Techniques and Applications of Plasma Chemistry (Hollahan, J. R., Bell, A. T. ed.), New York, Wiley — Interscience, 1974
37. Rodriguez, F.: Principles of Polymer Chemistry, New York, McGraw Hill, 1970
38. Vinzant, J. M., Shen, M., Bell, A. T., in: Plasma Polymerization — ACS Symp. Ser. 108, (Shen, M., Bell, A. T. ed.). Washington, D.C., Amer. Chem. Soc., 1979

Elementary Processes at Solid Surfaces Immersed in Low Pressure Plasmas

Harold F. Winters

IBM Research Laboratory, San Jose, California 95193
and
Physics Department, University of Odense, 5000 Odense, Denmark

Table of Contents

1 Introduction

Plasmas are being used for a wide and rapidly expanding variety of applications. These cover the entire range from very hot plasmas used to induce thermonuclear fusion reactions to cold plasmas which are used to deposit films or modify surfaces. For example, sputtering in a glow discharge environment has been used to produce a wide veriety of materials in thin film form[1]. Plasma-assisted chemical vapour deposition (CVD) has also been extensively used and the number of applications is rapidly increasing[2]. Plasma-assisted etching is now being introduced as a major tool in the manufacture of semiconductor devices[3]. Moreover, surface modification by exposure to a controlled plasma also has a number of applications[4].

The rapid increase in the use of plasmas has been accompanied by a large increase in the understanding of surface phenomena. Therefore, it is appropriate at this time to review the relationship between phenomena observed in a glow discharge and processes which are known to occur at solid surfaces.

Plasma environments are very complicated and many of the surface processes which occur in them have not been well studied. Moreover, investigators often are not able to define the basic parameters which are important in a given plasma situation. For example, the flux and energy of ions and electrons impinging upon a given surface are usually undefined. Furthermore, the types of neutral species and their concentrations in the gas phase are usually unknown. The field is in its infancy. Therefore, the type of knowledge most useful to investigators is often embodied in the following questions. First, does a given process occur? Secondly, what is a reasonable estimate for its rate? The organization of this paper is designed to answer these questions for the surface processes which the author considers to be most important in a glow discharge.

In order to answer these questions, three topics will be addressed for each surface process discussed. First, when the theory for a given process is available,

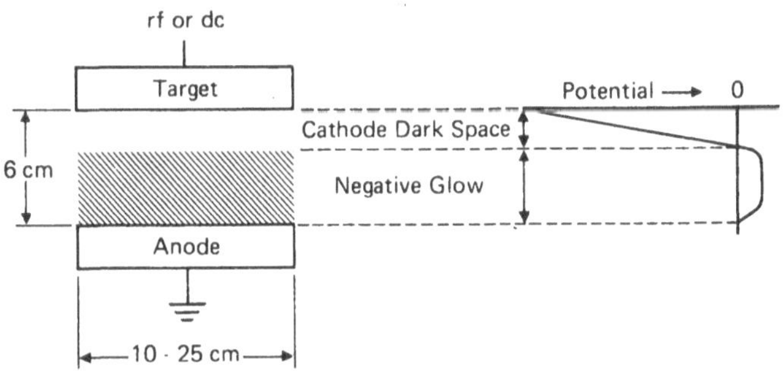

Fig. 1. Schematic of hardware used to generate glow discharge. Typical parameters: target diameter 10–50 cm, target-anode distance (3–30 cm), target current (0.2–2 mA/cm^2), target voltage — dc (−1000 to −5000 V), target voltage — rf (−100 to −1000 V), pressure (10^{-2} — 1 Torr), electron density in negative glow — (in the region of $10^{10}/cm^3$). dc voltage should be applied directly to the target while rf is applied through a capacitor. This type of system is usually operated in what is technically called the "Abnormal Glow Region." In this paper, examples of surface processes which occur in a glow discharge will be discussed with reference to this figure

it will be summarized. Secondly, representative experimental data will be presented which has been selected so as to allow estimates of appropriate rates. Thirdly, one or two examples of the importance of this surface process in a plasma situation will be presented.

The enormous quantity of literature and the large number of topics preclude a comprehensive review in the limited space available. Therefore, the author has chosen to limit discussion to that portion of the literature with which he is familiar and which conveys the desired information.

The illustrations, which relate glow-discharge observations to known surface processes, will be discussed in relation to the hardware and electrical configuration described schematically in Fig. 1. This type of hardware is widely used for sputtering, CVD, and etching. The application of the ideas presented in this paper to discharges generated in other types of hardware is straightforward but will be left to the reader.

2 The Interaction of Ions with Surfaces

In understanding the fate of an ion or excited atom as it approaches a metal surface, it is convenient to distinguish processes which occur relatively far from the surface where the potential energy (i.e., the electronic excitation energy) of the particle dominates from those processes which require a much closer approach or actual penetration of the lattice and therefore more directly involve its kinetic energy. Most evidence points to the conclusion that neutralization of slow ions and the resulting secondary-electron emission arise from processes falling into the first of these categories while momentum-transfer, cascade processes such as sputtering fall into the second category. Therefore, neutralization and secondary-electron emission are discussed in Sect. 2.1 and cascade effects in Sect. 2.2.

2.1 Neutralization and Secondary-Electron Emission

2.1.1 Basic Potential (Mechanisms Ejection)

Much of the experimental and theoretical information, which has lead to the understanding of ion neutralization at metal surfaces, has been provided by Hagstrum and coworkers[5-8]. Probst[9,10] has also made significant contributions to this field. The reader is referred to these original papers along with two recent reviews[11,12] for a more detailed discussion of this subject. The summary presented in this section generally follows the treatment of Hagstrum.

There are three possible mechanisms whereby an excited atom or ion can undergo an electronic transition near a metal surface: (1) de-excitation involving the emission of radiation, (2) de-excitation involving a two electron Auger process, and (3) a resonance process whereby an electron is transferred from the metal to an equivalent energy level in the ion or a similar transition where the electron goes from the ion to the metal. However, Schekhter[13] has shown that the probability for neutralization of an ion near a metal surface accompanied by radiation is very small ($\sim 5 \times 10^{-7}$). Moreover, the probability of any radiative process is expected to be small since the lifetime for radiation ($\sim 10^{-8}$ sec) is long when compared to the time that even a thermal particle spends within a few Angstroms of the surface ($\sim 10^{-12}$ sec). Therefore, electronic transitions near a metal surface must be dominated by Auger or resonance type processes which occur in about 10^{-14}–10^{-16} seconds.

The short lifetime of an excited species near a metal surface is most clearly demonstrated by the data in Fig. 2 and Table 1 which shows that the probability of an ion being neutralized upon collision with a surface is greater than 0.99 for a number of ions over a range of kinetic energies[14]. Neutralization happens with high probability even though the time that the ion spends near the surface is very short. The time that a 100 eV He^+ ion, for example, spends within 5 Å of the surface is $\sim 10^{-13}$–10^{-14} seconds and yet the probability of neutralization is 0.9983.

The basic physical processes involved in Auger neutralization are illustrated in Fig. 3. One electron tunnels into the ground state of the ion from the level in the metal labeled 1 while a second electron is simultaneously excited from

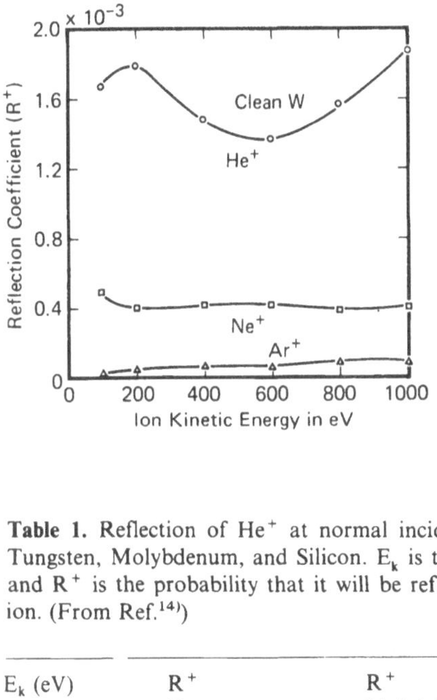

Fig. 2. Reflection of He^+, Ne^+, and Ar^+ from clean surfaces of tungsten. R^+ is the probability that the incident ion will be reflected away from the surface as an ion. (From Ref.[14])

Table 1. Reflection of He^+ at normal incidence from clean surfaces of Tungsten, Molybdenum, and Silicon. E_k is the energy of the incident ion and R^+ is the probability that it will be reflected from the surface as an ion. (From Ref.[14])

E_k (eV)	R^+ Tungsten	R^+ Molybdenum	R^+ Silicon
100	0.0017		
200	0.0018	0.0008	0.000024
400	0.0015		0.000038
600	0.0014	0.0009	0.00012
800	0.0016		0.0015
1000	0.0019	0.0010	0.00020

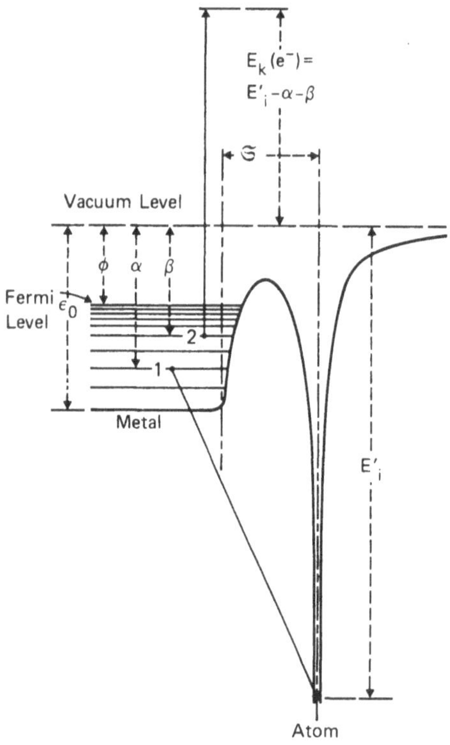

Fig. 3. Auger neutralization of an ion at a metal surface presented schematically. $E_k(e^-)$ is the kinetic energy of an electron observed outside the metal. ϕ is the work function and E_i' the ionization energy. ($E_i' < E_i$ where E_i is the energy needed to ionize an atom in free space.) $\mathfrak{S}$ is the distance of the ion from the surface. This figure is similar to one originally published in Ref.[5)]

another level labeled 2. The energy, which is released by the electron which tunnels into the ground state of the ion, is gained by the electron which is excited. The excited electron may leave the solid and be collected if it has sufficient momentum perpendicular to the surface. For most systems, this Auger process is the primary mechanism whereby low energy ions cause the emission of secondary electrons.

The energy level variations occurring for an atom or an ion near a surface is depicted by the potential energy diagram shown in Fig. 4. The initial state potential energy, which is specified as $He^+ + ne_w^-$ in this diagram, is a function of the distance '$\mathfrak{S}$' of the ion from the surface. At large distances this variation is a direct consequence of the image potential which is given by

$$V_i = -\frac{3.6\ (\mathrm{eV})}{\mathfrak{S}\ (\text{Å})},$$

while at smaller distances a repulsive interaction becomes dominant. One possible final state, which consists of a He atom, an energetic secondary electron, and the absence of two electrons at the Fermi level of the tungsten, is specified by $He + e^- + (n-2)\,e_w^-$ in Fig. 4. A large number of other possible final states with less potential energy are not shown in this diagram.

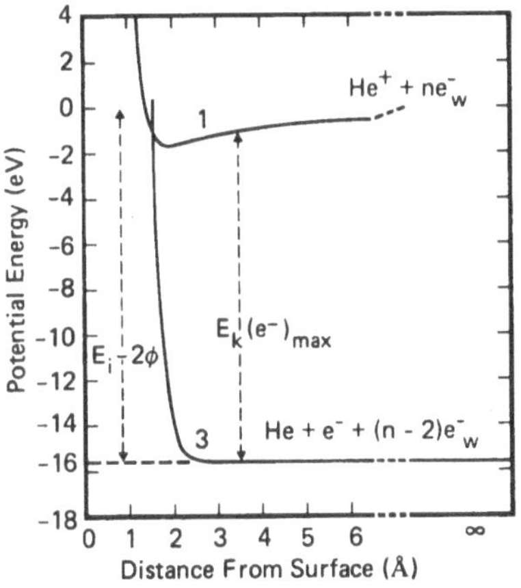

Fig. 4. Potential energy versus distance from the surface. Data is appropriate for He^+ and tungsten. E_i is the ionization potential for helium and ϕ is the work function of tungsten. $E_k(e^-)$ is the kinetic energy of an emitted secondary electron. The symbol $He^+ + ne_w^-$ implies a system composed of an helium ion and n conduction electrons in tungsten. The lower potential curve results from an Auger neutralization process where both electrons were originally at the Fermi level. (The figure is similar to one published in Ref.[5])

The maximum kinetic energy of electrons generated in an Auger process is predicted to be $E_i - 2\phi$ (see Fig. 4) while in general the kinetic energy would be $E_i' - \alpha - \beta$ (see Fig. 3). Experimental electron energy distributions (some of which are shown in Fig. 5) agree quite well with predictions about the maximum kinetic energy of ejected electrons ($E_i - 2\phi$) and also with other details of the distribution itself. In fact, ion-neutralization-spectroscopy, which is based upon the theory described above, yields band structure information which agrees with data obtained using other techniques such as ultra-violet photoelectron spectroscopy, UPS, and X-ray photoelectron spectroscopy, XPS.[11]

The secondary-electron yield is almost independent of energy (see Fig. 6), but depends strongly upon the ionization potential of the bombarding ion. Figure 6 shows that the yield for He^+ is ~ 0.24 while for Xe^+, it is less than 0.02.

Whereas most of the experimental data for ion-neutralization has been obtained using noble gases, there is strong evidence that neutralization of molecular ions

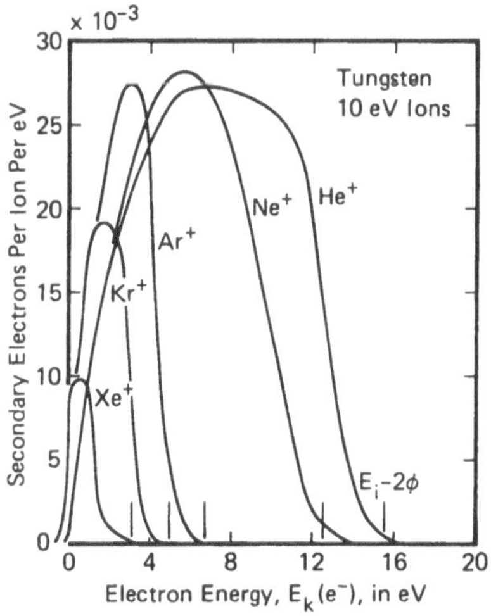

Fig. 5. Energy distribution functions for electrons ejected from tungsten by 10 eV noble gas ions. Verticale lines on the abscissa indicate the energies $E_i - 2\phi$. (From Ref.[15])

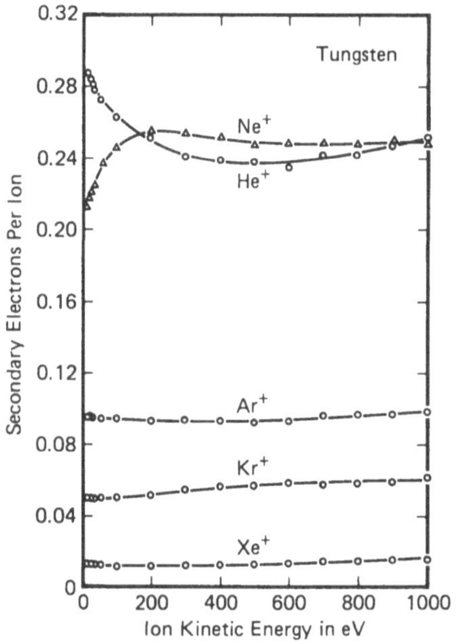

Fig. 6. Total electron yield versus ion kinetic energy for singly charged ions of the noble gases incident on atomically clean tungsten. Ions are all in the ground state. (From Ref.[15])

also involves an Auger process. Probst and Luscher[9] have investigated ion-induced secondary-electron emission for H_2^+, O_2^+, and N_2^+ on atomically clean tungsten. The secondary-electron yields were significantly smaller than would have been observed for an equivalent noble gas ion, but this behavior was attributed to the fact that the reflected neutral molecule left the surface in a vibrationally excited state. The other characteristics of the distributions were indicative of an Auger neutralization process.

In addition to Auger neutralization, resonance tunneling and Auger de-excitation are important processes at solid surfaces. These processes are described in Figs. 7a and 7b, respectively. In the transition labeled 1 (Fig. 7a), an electron from the metal tunnels into a level of the same energy in the ion leaving a neutral but excited atom. For the transition labeled 2, an electron from a metastable atom tunnels into an empty level in the metal leaving an ion. The resulting ion is subsequently neutralized by an Auger transition which can in turn produce a secondary electron as described above. In the Auger de-excitation process of Fig. 7b, an electron in an excited state of the atom is either ejected or drops to the ground state while a metal electron performs the other transition.

Whereas most of the experimental evidence demonstrating the dominance of the above mentioned mechanisms has been obtained for metals and semiconductors, it is our opinion that similar processes are likely to occur at insulator surfaces. Speculation about the influence of these electronic transitions on chemical reactions occuring at the surface will be presented in Sect. 4.3.

The implications of these results for plasma situations are quite clear. Ions and long-lived, electronically-excited molecules (or atoms) will be de-excited at walls and fixtures with high probability. These de-excitation processes will induce

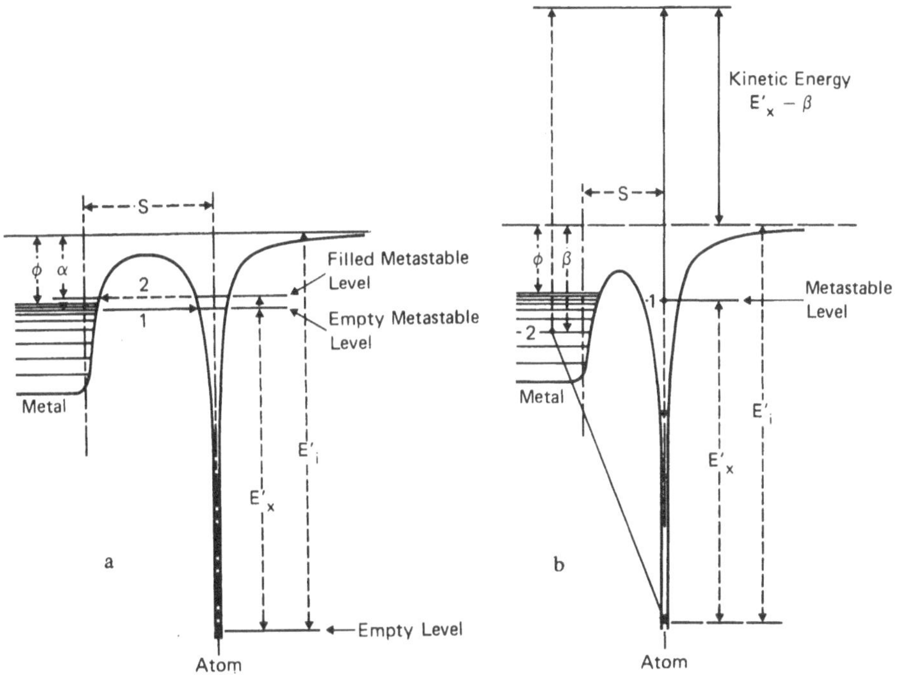

Fig. 7.a. Schematic diagram illustrating resonance neutralization of an ion (transition 1) or resonance ionization of an excited atom (transition 2) at a metal surface. Transition 1 can occur only at energy levels which are filled inside the metal and transition 2 at levels which are empty. (A similar figure was published in Ref. 5). **b.** Schematic diagram illustrating Auger de-excitation of an excited atom at a metal surface. The exchange transition is indicated by the full lines, while, the process not involving electron exchange between the metal and the atom is indicated by the dashed lines. (A similar figure was published in Ref.[5])

concentration gradients of the affected species within the plasma. Moreover, the de-excited neutrals which are reflected back from the surface may retain a significant fraction of their kinetic energy. They may also gain vibrational energy as a consequence of the de-excitation process. Therefore, surfaces in a plasma are often sources of energetic neutral particles and vibrationally excited molecules. (See Sect. 2.1.3 for more discussion of this subject.)

2.1.2 Basic Mechanisms (Kinetic Ejection)

For small kinetic energies, most of the secondary electrons are generated as consequence of the potential energy of the impinging ion (potential ejection); however, kinetic ejection becomes dominant as the kinetic energy is increased. The difference between potential and kinetic ejection is clearly established by the data of Medved et al.[16] (see Fig. 8), who compared the yield of secondary electrons from molybdenum induced by Ar^+ bombardment (potential plus kinetic ejection) with the yield obtained under bombardment by neutral argon atoms of

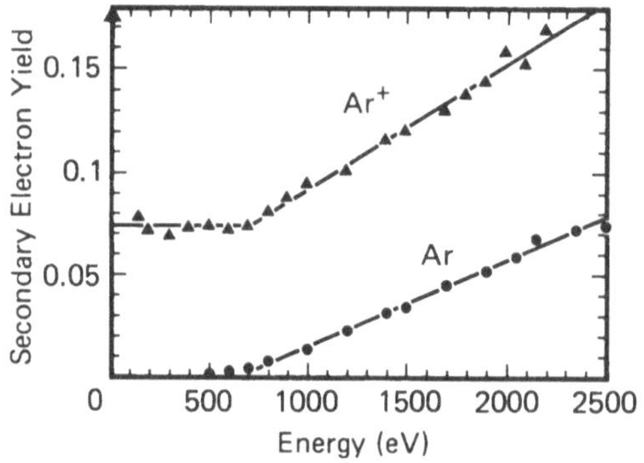

Fig. 8. Secondary-electron yields as a function of kinetic energy for Ar^+ and neutral argon atoms. (Data from Ref.[16])

the same kinetic energy (kinetic ejection only). The yields, to a first approximation, are identical except for a vertical displacement in the ion yields which is equal to the magnitude of potential emission. The potential emission yield is almost independent of energy while kinetic emission has a threshold at about 500 eV and increases linearly with energy thereafter. This type of behavior is also substantiated by the data of Arifov and Rakhimov[17] shown in Fig. 9. They made a thorough study of the secondary-emission coefficients in the 0.1–10 keV range using noble gas ions (potential plus kinetic ejection) and their nearest alkali ion counterparts (kinetic ejection only), i.e., Ne—Na, Ar—K, Kr—Rb. (Because of their small ionization potential, alkali metals do not produce potential emission of secondary electrons.) The pairs of yield curves almost coincide after a constant value equal to the estimated potential emission yield has been substracted from the noble gas results.

Whereas a theoretical framework for understanding potential emission of secondary electrons is well established, a similar quantitative framework for kinetic ejection must await future theoretical development. Nevertheless, kinetic ejection

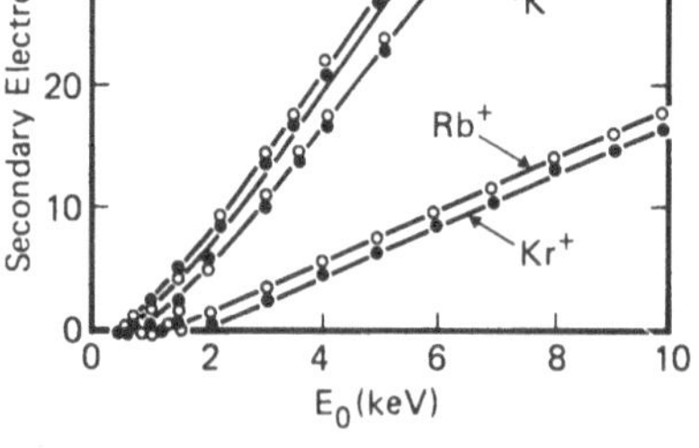

Fig. 9. Relative secondary-electron yield as a function of ion energy for Ne^+—Na^+, Ar^+—K^+, and Kr^+—Rb^+. A constant quantity equal to the estimated potential-secondary-emission coefficient has been substracted from the raw noble gas data. (From Ref.[17])

can be discussed on a semi-quantitative basis. A kinetic ejection yield calculation can be divided conceptually into a number of steps:

1) to determine the energy distribution of energetic electrons generated by an ion (or atom) of given kinetic energy and given position in the lattice,
2) to determine the probability that an electron of given energy and position will ultimately leave the solid, and
3) to determine the energy dissipated and trajectory of ions within the lattice.

As pointed out by Dietz and Sheffield[18], recent developments in the theory of inner shell vacancy production in ion—atom collisions add significant new knowledge about excitation processes which can lead to secondary-electron emission. In particular, Fano and Lichten[19], and Lichten[20] have applied molecular-orbital theory and the electron promotion mechanism to inelastic collisions of like atoms. Barat and Lichten[21] have extended the model to include unlike atoms. In their model, the electron shells inter-penetrate and a quasimolecule is formed during a collision process. Due to various mechanisms including the Pauli exclusion principle and curve crossings, electron promotion takes place and multiple electronic excitation can be generated in both collision partners. After they separate, the collision partners relax back to their ground states either by photon emission or by Auger electron emission. Garcia et al.[22] have reviewed work relating to inner shell vacancy production in ion—atom collisions while Burhop and Assad[23] have reviewed the Auger effect. In our opinion, a quantitative understanding of the kinetic ejection of secondary electrons will require a detailed understanding of these collision processes.

However, a much coarser approach can yield significant results. Several authors, for example, have attempted to relate the kinetic-secondary-electron coefficient to the stopping-power relationships used for the analysis of atomic ion ranges in solids. A very lucid paper on this approach has been presented by Beuhler and Friedman[24] and their perspective will be summarized in the following paragraphs.

Stopping power formulations are usually divided into two types. The first $(dE/dx)_e$ describes the energy loss per unit pathlength due to electronic excitation within the solid target while the second $(dE/dx)_n$ describes the energy loss due to collisions with target atoms. The total stopping power, $(dE/dx)_{tot}$, is the sum of the electronic and nuclear stopping powers and is determined in ion range measurements. In general, both $(dE/dx)_e$ and $(dE/dx)_n$ are easily calculable within the velocity range of most ion penetration measurements and agree with the measured values within a factor of two or better. However, in the low energy range, which is of interest to most investigators in plasma science, there is considerable uncertainty about the magnitude of these quantities.

Beuhler and Friedman[24] have developed a model, which agrees quite well with experiment, by assuming that the kinetic-secondary-electron coefficient, γ, is proportional to $(dE/dx)_e$. The proportionality of γ to $(dE/dx)_e$ has also been recognized by a number of other investigators. For example, at ion velocities less than 10^8 cm/sec, Schram et al.[25] noted "that there is an apparent analogy" between γ and $(dE/dx)_e$ while Dietz and Sheffield[18] concluded that γ should be proproportional to the amount of electronic excitation that an ion undergoes in the target. At high ion velocity, Sternglass[26] also used an assumption equivalent to $\gamma \propto (dE/dx)_e$.

The three pieces of information needed for a kinetic-secondary-electron yield calculation can now be related to the model of Beuhler and Friedman[24]:

1) The generation of energetic electrons within the solid is directly proportional to $(dE/dx)_e$.
2) The energy of the ion at a position within the target is determined by both $(dE/dx)_e$ and $(dE/dx)_n$.
3) The escape probability for electrons from the target is related exponentially to their depth.

The Beuhler-Friedman model becomes essentially identical to the earlier Sternglass model at high velocities. It is also quite similar to the well known model of Parilis and Kishinerskii[27] with the major difference being the method used to calculate the number of energetic electrons produced within the target.

At this time a few comments about the model of Parilis et al. is appropriate. Their theory predicts that no kinetic emission can occur up to a certain threshold energy; the yield then increases linearly with energy until at higher energy the slope again changes and the yield finally becomes linear with velocity. The majority of experimental results appear to confirm these trends suggesting that there is some truth in the basic assumptions.

2.1.3 The Influence of Ion-Induced Secondary-Electron Emission on Plasmas

Ion-induced emission of secondary electrons produces a variety of phenomena in plasma environments. Three of these phenomena will be discussed briefly for illustrative purposes. Consider, for example, the glow discharge described in Fig. 1. Ions are colliding with the target thus producing sputtered particles and secondary electrons. Much of the sputtered material condenses on the anode producing a thin film of target material. In addition, electrons are accelerated across the cathode dark space and collide at near normal incidence with the film growing on the anode. This electron bombardment produces significant heating, modifies film properties, causes electron-induced chemical reactions, and produces more secondary electrons which are accelerated back across the sheath into the plasma. The heating of the anode by secondary electrons was nicely demonstrated in the work of Ball[28] who found that up to 40% of the total energy coupled into the discharge appeared in the form of heat at the anode. Electron induced chemical reactions will not be discussed in this section but the reader is referred to Section 3.2 for a discussion of this subject. However, it should be remembered that chemical reactions occur in the presence of electron bombardment which would not normally occur in its absence.

In many situations, electrons leaving the target are not deflected substantially nor do they lose significant energy before reaching the anode. This conclusion is substantiated independently by the work of Chapman et al.[29] and Ball[28] and also explains many observations made by the author. It is also consistent with theoretical expectations since the inelastic collisions which do occur scatter electrons primarily in the forward direction[30]. Therefore several collisions on the average are needed to substantially deflect an electron. In this regard, consider a system of the type shown in Fig. 1 where the pressure is 5×10^{-2} Torr, the anode-target distance is 10 cm, and the gas has an ionization cross section of 2×10^{-16} cm^2 (see Ref. 32 for a compilation of cross Sect. for a variety of gases). A 3000 eV electron then will suffer ~4 collisions and lose about ~30 eV per ionization event[31]. If one assumes a secondary-electron coefficient of 0.1 (see Fig. 6), then each ion hitting the target will generate ~0.4 new ions as a consequence of the secondary electron traversing the discharge. This does not appear to be the dominant source of ions generated in the plasma but it is significant and probably greater than 10%.

In summary, the importance of ion-induced secondary-electron emission in plasma environments is demonstrated by, among other things, the fact that it produces a substantial fraction of the ions, generates significant heating of surrounding surfaces, and modifies the properties of deposited films.

2.2 Phenomena Induced by Momentum-Transfer

2.2.1 General Considerations

When an ion approaches a surface, it becomes neutralized and then begins to lose energy to the lattice. The means by which energy is lost is usually divided into two parts[33]; electronic collisions and nuclear collisions. The first process involves the interaction of the ion with the lattice electrons resulting in excitation and ionization. To a first approximation, the rate of energy loss to the lattice electrons is proportional to the ion velocity[33,34]. Except for light ions (H^+, He^+, etc.), the electronic energy loss is usually much smaller than the nuclear loss for the range of energies generally encountered in a glow discharge; therefore, electronic stopping will not be discussed from a theoretical point of view in this paper.

The so-called "nuclear loss" is that which results from collisions between the incident particle and the lattice atoms i.e. the deposited energy leads to the motion of lattice atoms. Many phenomena observed in glow discharges are related to collisions of this type. The first question, which must be considered, is whether the incident particle interacts with single surface atoms or with the lattice as a whole. Most investigators now believe that the incident ion can be assumed to interact with only one target atom at a time, and that the period of strong interaction is small when compared with the time required to travel between one atom and the next. The path of the incident ion is therefore determined by a series of binary collisions with the lattice atoms. The change in path caused by a given encounter is determined by conservation of energy and momentum, i.e., classical dynamics.

The importance of the binary collision approximation in the development of an understanding of ion-surface collision phenomena would be difficult to over-emphasize. Many successful theories[35,33,36–38], which describe a wide range of phenomena, are based on this approximation. Moreover, binary collisions can be demonstrated experimentally in the backscattering of ions from surfaces (see Fig. 10). For a single binary collision, the quantity $(E - T)/E$ is independent of E for constant Φ where E is the kinetic energy of the incident ion, T is the energy transferred during the collision, and Φ is the scattering angle (see Fig. 11). The data of Heiland and Taglaur[39], which is shown in Fig. 10, demonstrates that this quantity is indeed

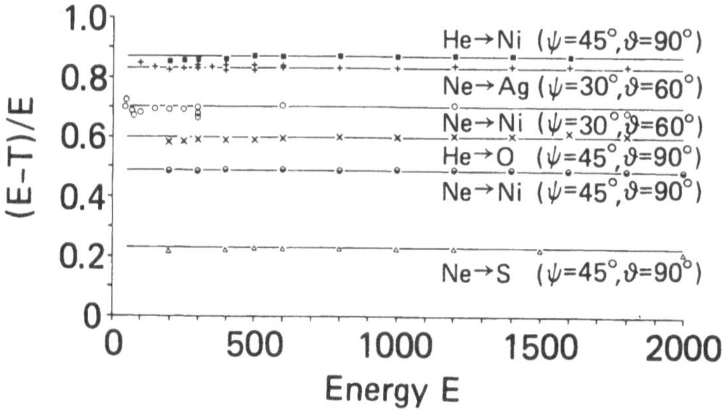

Fig. 10. Fractional energy loss (E — T)/E vs. primary ion energy. The scattering angle, θ, and the angle of incidence, Ψ, are constant. (Data from Ref.[39])

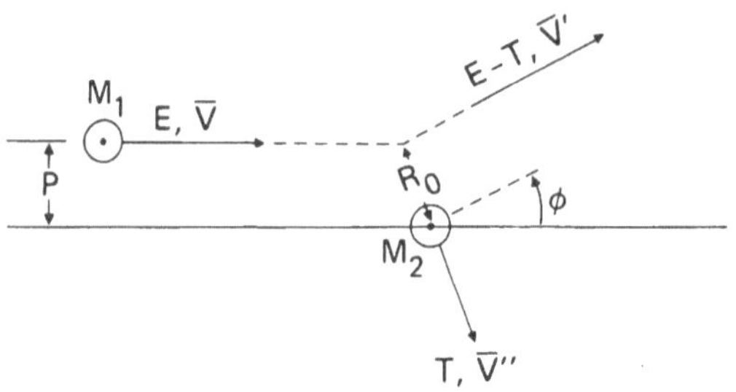

Fig. 11. Scattering of two particles as viewed from the laboratory system. E is the initial ion energy and $\bar{V}$ the initial velocity. M_1 and M_2 are masses

independent of energy down to at least 100 eV. This result is consistent with the computer simulations of Karpuzov and Yurasova[40a] who investigated the reflection of 50–500 eV argon ions from a copper crystal and concluded that ion reflection is adequately described by the binary collision approximation. Similar conclusions can also be drawn from the simulation studies of Poelsema et al.[40].

Robinson and Torrens[41] have discussed the limitation of the binary-collision approximation. They concluded, for example, that it was likely to fail at energies below 9 eV for collisions between copper atoms in a copper lattice and below 33 eV for gold atoms in a gold lattice. Therefore, phenomena which depend sensitively upon motions of particles with very low energy are likely to be described only qualitatively by the binary collision model.

Even though the binary collision model is very useful, it is still an approximation to the real situation and a more detailed understanding probably requires computer simulation studies such as those pioneered by Harrison[42,43].

When an ion collides with a solid surface, the collisions between quasi-independent particles result in a cascade of moving target atoms — hence the term collision cascade. The consequences of the collision cascade include: sputtering, radiation enhanced diffusion, backscattering of the incident ion, ion-induced chemical reactions, production of lattice damage, etc. The initial collision involves an energetic ion (or atom) and a stationary (neglecting thermal energy) target atom. Subsequent collisions may involve moving atoms colliding with stationary atoms or may involve the collision of moving atoms with each other. In the first instance, linear cascade theory — such as that developed by Sigmund to explain sputtering phenomena — is applicable. In the second instance (moving atoms colliding with moving atoms) an equilibrium description in which temperature is a meaningful concept might possibly become useful. Evidence for both extremes can be observed. The pertinent question in any given situation is: which dominates? It is the prejudice of the present author that linear cascade theory is a good approximation in many situations. Moreover, we seriously doubt that temperature ever becomes a meaningful concept. This rather extreme conclusion is based on the fact that energy is not confined within a given volume, and that the number of collisions are too few to produce an entire region which is in the thermodynamic equilibrium. However, it may prove to be *impossible* to describe phenomena, such as ion induced chemical reactions or radiation enhanced diffusion, by either linear cascade theory or by using the concept of temperature. These phenomena may be dominated by an intermediate situation where non-linear effects in the cascade are important. Moreover, even in situations where linear cascade theory is a useful approximation, non-linear effects are likely to be observed.

2.2.2 Theoretical Considerations

The theoretical ideas of Bohr[44] and more recently of Lindhard and co-workers[33] have been a major force leading to the understanding of ion-surface interactions. We will now summarize these ideas generally following the reasoning used by Lindhard[33,45,46] and Sigmund[35]. The key quantity is the specific energy loss, which is the average energy loss per unit path length, and is defined by

$$\frac{dE}{dx} = N \cdot S = N \int d\sigma(E, T) \cdot T$$

where N is the number of scattering centers per unit volume and S is the stopping cross section per center. $d\sigma(E, T)$ is the differential cross section for transfer of energy between T and $T + \delta T$ from the ion to the target atom. We will now derive a relationship for $\int T \cdot d\sigma(E, T)$.

Two-particle collisions are often adequately described using the ideas of classical dynamics. The quantity, $d\sigma$, can be formally obtained from the following well-known equations[47,48]:

$$T = T_m \sin^2 \frac{\theta}{2}; \qquad T_m = \frac{4M_1M_2}{(M_1 + M_2)^2} E, \tag{1}$$

$$\theta = \pi - 2P \int_{R_0}^{\infty} \frac{dR/R}{\left[1 - \frac{U(R)}{E_R} - \frac{P^2}{R^2}\right]^{1/2}}, \tag{2}$$

$$E_R = \frac{M_2E}{M_1 + M_2},$$

$$d\sigma(\theta) = -2\pi P \, dP, \tag{3}$$

where (see Fig. 11) M_1 and M_2 are the masses of the colliding atoms, E_R is the primary energy measured with respect to the center of mass, T the energy transfer (the energy of the recoiling atom), θ the center of mass scattering angle, P the impact parameter, R the internuclear distance, R_0 the distance of closest approach, and U(R) the interatomic potential. T_m is the maximum possible energy transfer and R_0 is the root of the equation[48]:

$$\left[1 - \frac{U(R_0)}{E_R}\right] R_0^2 = P^2 ,$$

dσ is formally obtained by inverting Equation 2 to obtain P as a function of θ. P and dP are then substituted into Equation 3 yielding dσ(θ) which is subsequently changed to dσ(T) using Equation 1. The input quantity needed to use this procedure is the interaction potential, U(R).

It is not absolutely necessary to have accurate interatomic potentials to perform reasonably good calculations because the many collisions involved tend to obscure the details of the interaction. This, together with the fact that accurate potentials are only known for a few systems makes the Thomas-Fermi approach quite attractive. The Thomas-Fermi statistical model assumes that the atomic potential V(r) varies slowly enough within an electron wavelength so that many electrons can be localized within a volume over which the potential changes by a fraction of itself. The electrons can then be treated by statistical mechanics and obey Fermi-Dirac statistics. In this approximation, the potential in the atom is given by:

$$V(r) = \frac{Ze^2}{r} \phi\left(\frac{r}{a}\right); \qquad a = 0.885\, a_0 Z^{-1/3} , \tag{4}$$

where Z is the atomic number and a_0 the Bohr radius, 0.529 Å. For a derivation of Equation 4, see Ref. 49. ϕ(r/a) is the Thomas-Fermi screening function shown in Figure 12.

It is convenient to describe the interatomic potential, U(R), with the same functional form as Equation 4. This has been accomplished by Bohr[44] who estimated the interaction energy between two atoms by the formula:

$$U(R) \cong \frac{Z_1 Z_2 e^2}{R} \exp\left(-\frac{R}{a}\right),$$

where R is the internuclear distance and exp (−R/a) is the screening function. Subsequent authors usually represented their interaction potentials in the same form but with modified screening functions. Firsov[50] showed that, within the limits of accuracy of the Thomas-Fermi statistical model, the interaction between atoms at a distance less than 10^{-8} cm could be described by the potential:

$$U(R) = \frac{Z_1 Z_2 e^2}{R} \phi\left(\frac{R}{a_F}\right), \tag{5}$$

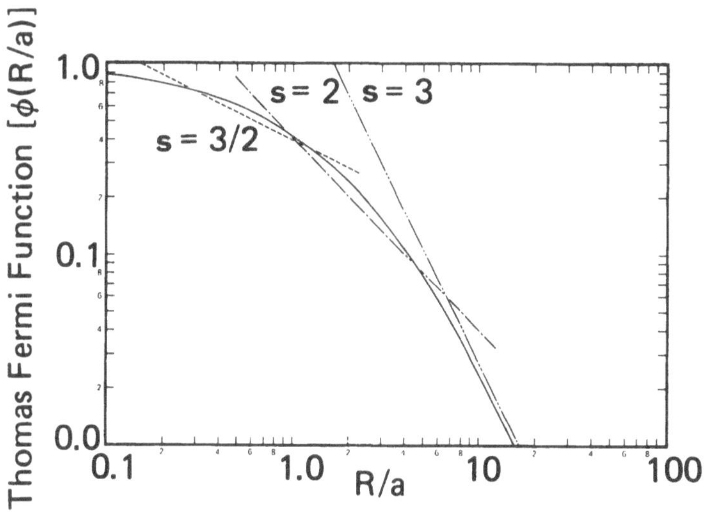

Fig. 12. Thomas-Fermi screening function, $\phi(R/a)$ (see Equation 4), for neutral atoms (———) and power approximations (– – –) from Equation 7. Values of $\phi(R/a)$ are from Ref.[51]. Constants used in Equation 7 are: $k_{1.5} = 0.591$, $k_2 = 0.833$, $k_3 = 2.75$. See Equation 7 for definition of k

where $\phi(R/a_F)$ is the Thomas-Fermi screening function shown in Fig. 12 and

$$a_F = 0.8853 a_0 (Z_1^{1/2} + Z_2^{1/2})^{-2/3} .$$

Lindhard et al.[45] preferred a screening radius:

$$a_L = 0.8853 a_0 [Z_1^{2/3} + Z_2^{2/3}]^{-1/2} ,$$

for the same functional form of Equation 5. The two screening radii are numerically equal within the accuracy of the Thomas-Fermi approach. In subsequent sections we will use a_L and refer to it as "a".

Equation 5 is often used to decribe the interaction between the incoming ion and the target atoms. The interaction between two target atoms generally occurs at low energy where the Thomas-Fermi potential overestimates the interaction. Under this situation a Born-Mayer potential is more appropriate[35], i.e.:

$$U(R) = A e^{-bR} , \tag{6}$$

where A and b are constants. Typical values for A and b have been tabulated by Abrahamson[52].

An especially useful approximation for the Thomas-Fermi potential has been developed by Lindhard[45] and co-workers where the screening function is assumed to have the form:

$$\phi\left(\frac{R}{a}\right) = \frac{k_s}{a}\left(\frac{a}{R}\right)^{s-1} , \tag{7}$$

where k_s and s are constants. U(R) then becomes:

$$U(R) = \frac{k_s}{s} \frac{Z_1 Z_2 e^2 a^{(s-1)}}{R^s}. \tag{8}$$

Figure 12 shows that Equation 7 reasonably approximates the screening function over limited energy ranges. The inverse power approximation made in Equation 7 is quite attractive since it allows Equation 2 to be integrated in closed form for several values of s[53].

The substitution of Equation 8 into Equation 2 followed by approximations (see Ref. 45) and integration leads to:

$$d\sigma(T) = CE^{-m}T^{-(m+1)}\,dT; \qquad m = \frac{1}{s}, \tag{9}$$

where

$$C = \frac{1}{2}\pi\lambda_m a^2 \left(\frac{M_1}{M_2}\right)^m \left(\frac{2Z_1Z_2e^2}{a}\right)^{2m},$$

and $\lambda_1 = 0.5$, $\lambda_{1/2} = 0.327$, $\lambda_{1/3} = 1.309$. m = 1 corresponds to Rutherford scattering, and Sigmund[15] has shown that m = 0 approximates scattering from a Born-Mayer[35] potential. For m = 0, $C_0 = (1/2)\,\pi\lambda_0 a^2$ where $\lambda_0 = 24$ and a = 0.219 Å. For m = 0, "a" is assumed to be independent of Z. Sigmund also suggests that for medium mass ions and atoms over most of the keV range, m = 1/2 is a fair approximation while in the lower keV and upper eV region, m = 1/3 should be adequate.

Equation 9 is an extremely useful description for the differential cross section and has been used extensively in a variety of applications. With the use of Equation 9, the nuclear stopping power becomes:

$$S_n(E) = \int_0^{T_m} T\,d\sigma = \frac{C}{1-m}\left[\frac{4M_1M_2}{(M_1+M_2)^2}\right]^{1-m} E^{1-2m}. \tag{10}$$

2.2.3 Reflection and Trapping of Incident Ions

2.2.3.1 Theory and Experiment

As mentioned earlier, many of the physical and mathematical concepts used by modern investigators to understand the interaction of energetic particles with solids are based on the original work of Lindhard[33,45,46] and Bohr[44]. Therefore, many similarities exist between the theories which describe sputtering[35,36], the range of ions in a solid[33], the generation of lattice damage[38], and the reflection and trapping of energetic particles[33,54]. We will briefly describe the concepts used to understand the trapping and reflection of energetic particles, and leave it to the reader to apply similar procedures to the phenomena of sputtering and ion ranges.

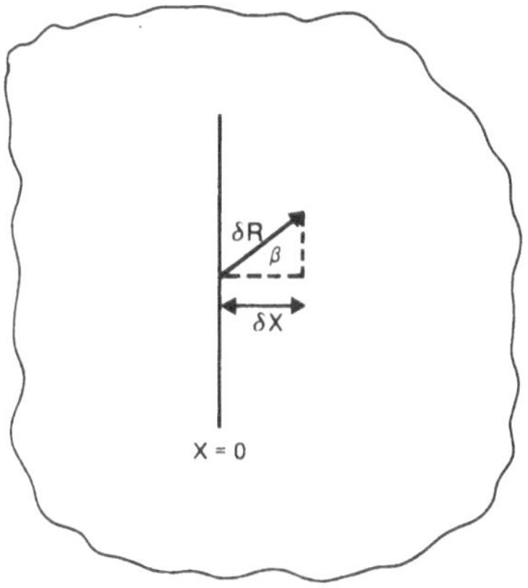

Fig. 13. Schematic drawing used to illustrate theoretical concepts for calculation of sputtering yields, and for calculation of reflection and trapping coefficients. The particle begins its motion at x = 0 in an infinite medium in a direction defined by the angle β

Suppose that an atom starts it motion with an energy E in an arbitrary direction from a plane $x = 0$ in an infinite lattice composed of atoms placed at random locations (see Fig. 13). The probability that an atom will come to rest at a distance "x" from the starting point is given by the range distribution, $F_R(x, E, \eta)$, where $\eta = \cos\beta$ and β is the initial angle between the beam and the x-direction. Then the trapping coefficient is given by

$$T_r = \int_0^\infty F_R(x, E, \eta)\,dx\,, \tag{11}$$

and the reflection coefficient is given by

$$R_e = \int_{-\infty}^{0} F_R(x, E, \eta)\,dx\,. \tag{12}$$

A derivation of an integral equation for F_R will now be presented which closely follows the procedure used by Lindhard et al.[33] and uses the nomenclature of Weisman and Sigmund[54]. A particle, which starts at $x = 0$, may or may not have suffered a collision after moving a distance $\delta R = \delta x/\cos\beta$; nevertheless, the final distribution will not have changed. Therefore, F_R based on initial conditions is equated to an expression for F_R based on conditions which prevail after the particle has traveled a distance δR.

There is a probability $N\,\delta R\,d\sigma$ for a collision specified by energy transfer "T" to the translational motion of the struck atom. N is the volume density of atoms, and one approximation for dσ is given by Equation 9. If a collision occurs, the particle has a probability $F(x - \delta x, E - T, \eta')$ of obtaining a total projected range, x. Therefore, $N\,\delta R\,d\sigma\,F(x - \delta x, E - T, \eta')$ is the contribution from this specified collision to the total probability for the projected range, x. When this term space is integrated over all collisions, the total contribution becomes:

$$N\,\delta R \int d\sigma\,F_R(x - \delta x, E - T, \eta')\,.$$

There is left a probability $1 - N\,\delta R \int d\sigma$ that no collision occurs. In this event, we clearly get a contribution $[1 - N\,\delta R \int d\sigma]\, F_R(x - \delta x, E, \eta)$ to the total probability for the projected range, x. Equating these two terms to the initial distribution yields:

$$F_R(x, E, \eta) = N\,\delta R \int d\sigma\, F_R(x - \delta x, E - T, \eta') + + [1 - N\,\delta R \int d\sigma]\, F_R(x - \delta x, E, \eta)\,. \tag{13}$$

Expansion of F_R in terms of δx followed by algebraic manipulation yields to first order:

$$-\eta \frac{\delta F_R}{\delta x} = N \int d\sigma [F_R(x, E, \eta) - F_R(x, E - T, \eta')]\,. \tag{14}$$

If electronic stopping is important this equation becomes[55]:

$$-\eta \frac{\delta F_R}{\delta x} = NSe \frac{\delta F_R}{\delta E} + N \int d\sigma (F_R - F_R')\,. \tag{15}$$

The reader is referred to Ref. 33 for solutions to this type of equation.

It should be remembered that F_R was derived on the basis of assuming an amorphous solid. The same is true for Sigmund's sputtering theory. However, it is the suspicion of the present author that the sequence of collisions, which are present in every cascade, tends to make the random assumption more realistic. Therefore, it is possible that this theory can be applied with some degree of confidence to polycrystalline materials. It cannot, of course, be applied to single crystal materials where chaneling and related phenomena occur.

In this model for reflection (Equation 12), $x = 0$ is merely a reference plane within the target where the ion (atom) motion starts, i.e., the target extends to infinity on both sides of this point. Some particles start with a velocity vector directed in the positive x direction; but, after losing their energy, they are located at negative values of x. These are the reflected particles. This model will underestimate the reflection coefficient; because, in the theoretical model an atom can pass through the hypothetical surface several times, while in practice it will be backscattered through the surface only once. Nevertheless, in many instances this model is quite good[37], while, in other situations an easily applied surface correction produces better results[56]. As a rule of thumb, electronic stopping (see Ref. 57 for magnitude estimates) and a surface correction are important for light ions while both can be neglected when heavier ions are being investigated.

These calculations should be quite good for light ions and for heavy ions in the keV range, but are probably not useful for heavy ions in the eV range. Figure 14 shows the calculated reflection coefficient as a function of energy for hydrogen. deuterium, tritium, and helium on niobium. The reflection coefficients are large, and increase with decreasing ion energy. Results from the computer simulation studies of Oen and Robinson[57] exhibit similar trends (see Fig. 15) as do the experimental results of Eckstein et al.[58]. However, the uncertainties in the absolute numbers may approach a factor of two.

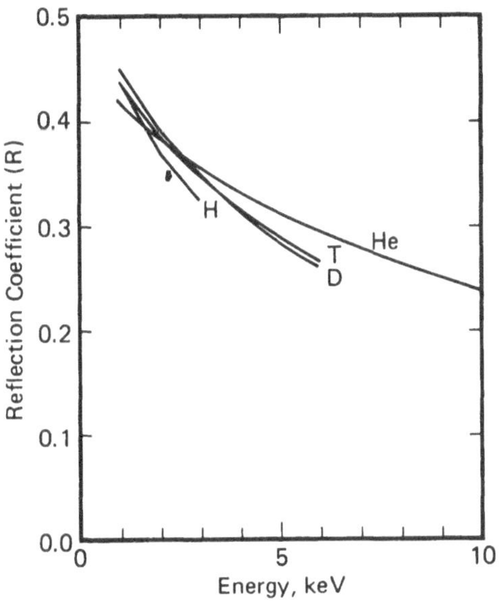

Fig. 14. Theoretical calculations of the reflection coefficients as a function of energy for H^+, D^+, T^+, and $^4He^+$ incident on Nb. Surface correction has been applied. (Data from Ref.[57])

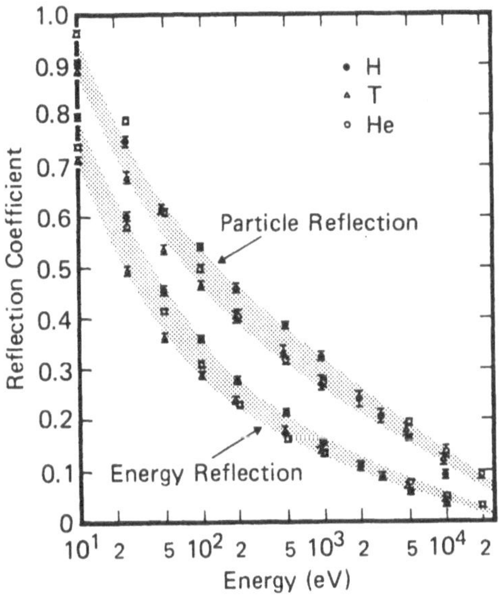

Fig. 15. Theoretical calculations of the reflection of energy and the reflection of particles for H, He, and T from copper. Calculations based on computer simulation. Normal incidence. R_E/R_N is the average fractional energy of a reflected particle. (From Ref.[57])

In the inverse power approximation (see Equations 5 and 7) and in the absence of electronic stopping, the reflection coefficient is independent of energy. It only depends on the angle of incidence, the mass ratio (M_2/M_1, see Fig. 11), and the value of m (see Equation 9 and following paragraph). Therefore, estimates for the reflection coefficients for medium to heavy ions in the keV range can be obtained from Figs. 16 and 17. The reflection coefficient increases for increasing values of M_2/M_1 and also for increasing angle of incidence.

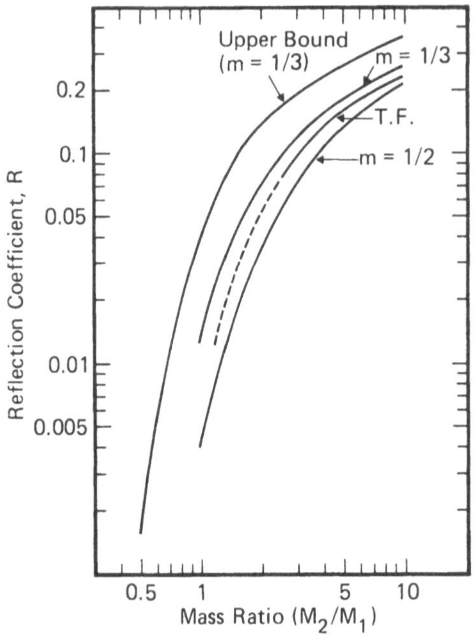

Fig. 16. Calculated values for the reflection coefficient as a function of mass ratio. M_2 is the mass of the target atom. Three curves, corresponding to the Thomas-Fermi (T.F.) and the two power approximations ($m = {}^1/_3$ and $m = {}^1/_2$) cases are shown. No surface correction has been applied. (From Ref.[37])

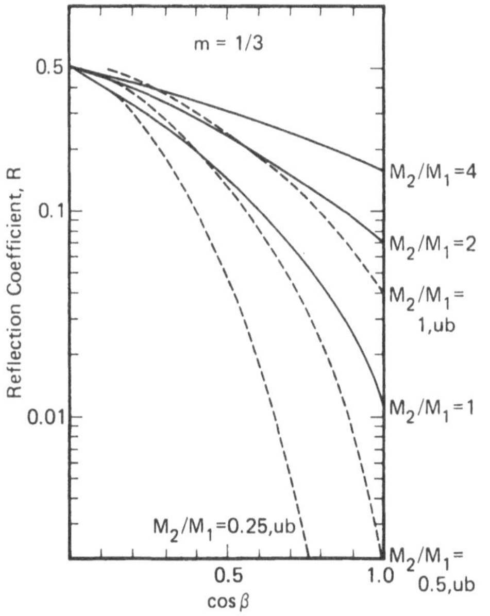

Fig. 17. Dependence of the calculated reflection coefficients on angle of incidence for 5 mass ratios (see Fig. 13). M_2 is the target mass. No surface correction was made. Note, a surface correction becomes quite important as cos β becomes small, i.e., when the angle of incidence becomes large. For mass ratios $M_2/M_1 \leq 1$, upper bounds are shown (dashed lines). (From Ref.[37])

Whereas, the theoretical values for the reflection coefficients shown in Figs. 14–16 are reasonable estimates for light ions and for heavy ions in the keV range, there is no adequate theory for ions in the eV range. There have, however, been some computer simulations which seem to correctly predict trends[59]. Moreover, excellent experimental work has been reported by Kornelsen and co-workers[60–62] for ions in this energy range.

Figure 18 shows the experimental trapping probability ($1 - R_e$) for the noble gas ions bombarding tungsten. For energies below approximately 50 eV, the trapping probabilities are negligibly small; while, for energies above 2000 eV, the sticking probabilities are all quite large.Atoms of large mass tend to have small trapping probabilities (large reflection coefficients) while atoms of small mass have larger trapping coefficients. Trends similar to those observed for tungsten appear to occur for a wide variety of materials[63-65].

Whereas, the trends are the same for different materials, subtle differences in the condition of the surface can substantially affect the absolute values. For example, the trapping probability for Ar^+ in nickel can be increased by a factor of two by adsorbing a layer of oxygen upon the surface[64].

Even though the trapping coefficients for the noble gases are quite large, the steady state concentration of gas in the surface region is usually small[62]. This is a direct consequence of the fact that ion bombardment very effectively releases previously trapped gas. For example, Kornelsen[61] has shown that the area affected by a single ion can exceed 10^{-13} cm^2. Noble gas atoms will not remain on a surface at room temperature; therefore, any process which allows the noble gas atoms to diffuse to the surface will also cause them to desorb into the gas phase. The release of previously trapped noble gas by ion bombardment is not a sputtering phenomena but is probably dominated by a radiation-enhanced diffusion-process. Therefore, the conceptual framework used to understand physical sputtering should not be used to describe the release of noble gas.

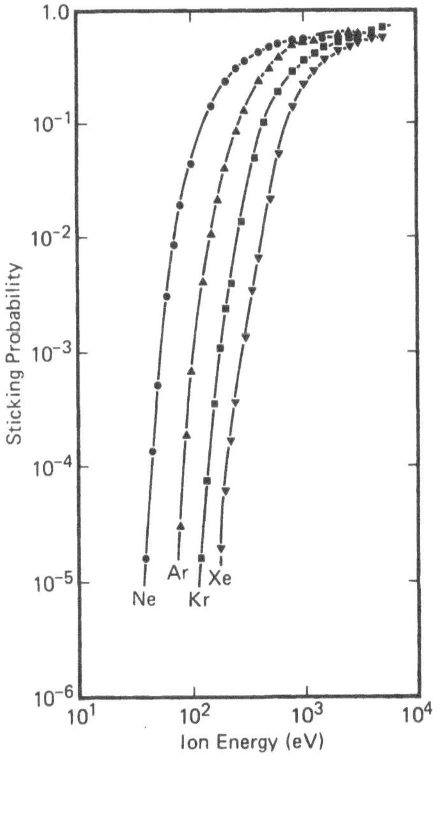

Fig. 18. The sticking probability of inert gas ions on tungsten as a function of ion energy. (From Ref.[62])

The trapping and reflection of chemically active gas is likely to exhibit behavior similar to the noble gases; however, this topic will be discussed in the section on ion-induced chemical reactions.

2.2.3.2 Reflection and Trapping of Particles in a Plasma Environment

Although many other situations could be discussed, the importance of the reflection and trapping of energetic particles in a plasma environment will be illustrated by discussing the glow-discharge, sputter-deposition of thin films in a system of the type shown in Fig. 1. Argon ions are accelerated into the target; and as a consequence, sputtered material is deposited upon the anode. In the arrangement of Winters and Kay[66)], ions with energies up to 3000 eV strike the target while the anode, which is at ground potential, is bombarded with ions whose energy is less than ~10 eV. (Note, because of charge exchange and other collisional processes, the average ion does not collide with the target with an energy equivalent to the difference between the plasma potential and the target potential[67)].) A small probe (~1 cm^2) in the anode plane could be biased with respect to the plasma in these experiments. Thus, the energy of the ions, which were bombarding the film growing on the probe, could be controlled. Figure 19 shows the argon content of the film as a function of bias voltage. The qualitative behavior of this data is consistent with information which has subsequently been published for a number of similar situations[68−71)].

Films deposited in a glow discharge always contain concentrations of gases which make up the ambient atmosphere. For argon, the concentrations may vary between 0.1–50 at. %, depending upon the situation.

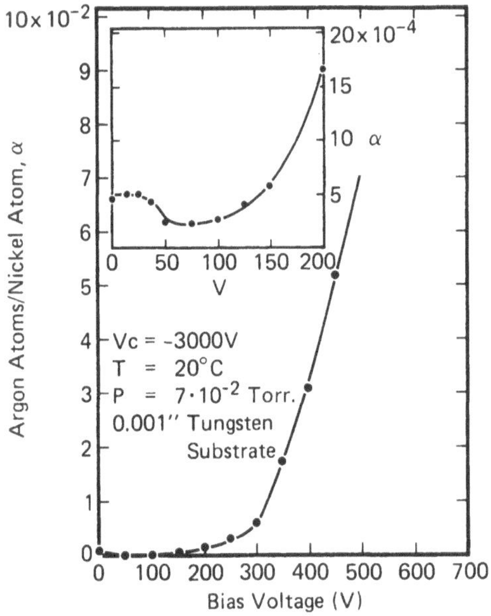

Fig. 19. The argon concentration of nickel films grown in a glow discharge vs. bias voltage. Evaporated films grown under similar conditions, but without a discharge, would not contain argon. (From Ref.[66)])

The interpretation of the data shown in Fig. 19 is straightforward when it is remembered that: (1) atoms with thermal energy are not incorporated into the film[66], (2) particles whose energy is less than approximately 50 eV are not trapped, and (3) particles whose kinetic energy is greater than 100 eV have a significant sticking probability. The increased argon concentration for bias voltages greater than ~100 V is due to ions which are accelerated across a thin sheath and into the growing film. This interpretation is quite consistent with the data shown in Fig. 18. In contrast, the argon trapped at zero bias is *not* due to ions which are accelerated across the sheath and into the film. The kinetic energy of these ions is much too small; therefore, they will not be trapped. However, some of the ions, which hit the target with kinetic energies of up to 3000 eV, are first neutralized and then reflected back into discharge with a large fraction of their initial kinetic energy. In the energy range between 100–600 eV, for example about one-half of the energy carried by the incident ions is transferred to a tungsten target and the other half is carried away by the reflected neutrals[72]. Some of these neutrals cross the discharge and become trapped in the growing film. This accounts for the argon concentration in sputtered films at zero bias voltage. The decrease in the argon concentration for bias voltages between 50–100 V is due to the fact that ions with this low energy are not significantly trapped, but, they do effectively release previously trapped gas as indicated by Kornelsen[61].

We have described the reflection and trapping of energetic particles which occur in one type of discharge environment; however, similar phenomena must occur in a wide variety of situations.

2.2.4 Sputtering

The ejection of material from a solid surface under bombardment by energetic ions (or neutrals) is known as sputtering. The sputtering yield, Γ, is defined as the number of target atoms ejected per incident ion. From the perspective of plasma science, it is most important to understand how the sputtering-yield-parameter behaves under a wide variety of situations. Consequently, the scope of this section will be limited to presenting a perspective on this subject. Review articles on sputtering have been published by Günthcrschulzc and Meyer[73], Wehner[74], Behrish[75], Kaminsky[76], Carter and Colligon[47], MacDonald[77], Tsong and Barber[78], Oechsner[79], and Winters[80]. McCracken[81] has recently published a paper on the interaction of ions with solid surfaces in which there is a long section on sputtering. In addition, Sigmund[82–84] has also published a seris of review articles on displacement damage which includes a chapter on sputtering. Because of the large number of review articles, the length of this section is minimized and the reader is referred to these articles for a more detailed discussion.

2.2.4.1 Sputtering-Theoretical

There have been three important theories in recent years on the sputtering of amorphous and polycrystalline solids. They are attributed to Sigmund[35], Thompson[36], and Brandt and Laubert[85]. The predictions about various aspects of sputtering often agree. Sigmund's theory, however, is the most general and latest and therefore will be discussed in this section.

A yield calculation consists of five steps:

1) To determine the differential cross section $d\sigma(E, T)$ for the transfer of energy between T and T + dT from the ion to the target atom and from one target atom to another. $d\sigma(E, T)$ is, for example, approximated by Equation 9 for an inverse power potential.
2) To determine the amount of energy deposited in the surface region. In many instances this is approximated by a quantity proportional to the nuclear stopping power (see Equation 10).
3) To convert this energy into the density of low energy recoil atoms.
4) To determine the number of recoil atoms which reach the very surface.
5) To select the fraction of these atoms which are able to overcome the surface binding forces and thus be emitted into the gas phase.

In Sigmund's theory, steps 2–4 were accomplished by deriving a differential equation for the sputtering yield based on methods similar to those outlined for reflection of energetic particles in Sect. 2.2.3 (see Ref. 35 for more detail). The sputtering yield from Sigmund's theory is given by

$$\Gamma(E, \eta) = 0.042\alpha \frac{S_n(E)}{U_0}. \tag{16}$$

The yield is essentially proportional to the amount of energy deposited in a thin layer near the surface (i.e., to the nuclear stopping power, $S_n(E)$) and inversely proportional to the binding energy (U_0) of the atom to the surface. The binding energy can be approximated by known sublimation energies[86]. α is a dimensionless quantity which depends upon the relative masses and the angle of incidence. Direct energy dependence in α drops out for power scattering. α as a function of M_2/M_1 is shown in Fig. 20.

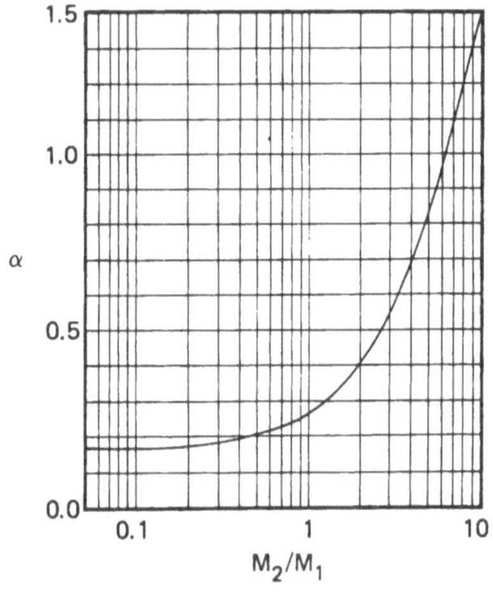

Fig. 20. Factor α in sputtering yield formula (Equation 16) calculated for power scattering and averaged between m = $^1/_3$ and m = $^1/_2$. (From Ref.[35])

For low energies (m = 0; E $\lesssim$ 1 keV), the yield becomes

$$\Gamma(E,\eta) = \frac{3}{4\pi^2}\,\alpha\,\frac{4M_1M_2}{(M_1 + M_2)^2}\,\frac{E}{U_0}\,, \tag{17}$$

where the value for $S_n(E)$ was obtained from equation 10. Equation 10 can also be used for keV energies and heavy-to-medium mass ions; however, the expression for the nuclear stopping power calculated by Lindhard, assuming a Thomas-Fermi interaction is often substituted, i.e.,

$$S_n(E) = 4\pi Z_1 Z_2 e^2 a \left[\frac{M_1}{M_1 + M_2}\right] s_n(\varepsilon)\,,$$

where $s_n(\varepsilon)$ is the reduced stopping power shown in Fig. 21. The sputtering yield from Equation 16 is, therefore,

$$\Gamma(E,\eta) = 3.56\alpha\,\frac{Z_1Z_2}{[Z_1^{2/3} + Z_2^{2/3}]^{1/2}}\,\frac{M_1}{M_1 + M_2}\,\frac{s_n(\varepsilon)}{U_0}\,, \tag{18}$$

where U_0 is in eV. The relationship between the ion energy E and the reduced energy ε is given by:

$$\varepsilon = \frac{aM_2E}{[Z_1Z_2e^2(M_1 + M_2)]}\,. \tag{19}$$

Figures 20 and 21 along with Equations 18 and 19 allow yield calculations for any ion-target combination.

The Sigmund theory, strictly speaking, is only valid for amorphous, monatomic targets and scattering using an inverse power potential. Moreover, it should be most accurate for medium mass ions in the keV range and relatively more inaccurate in

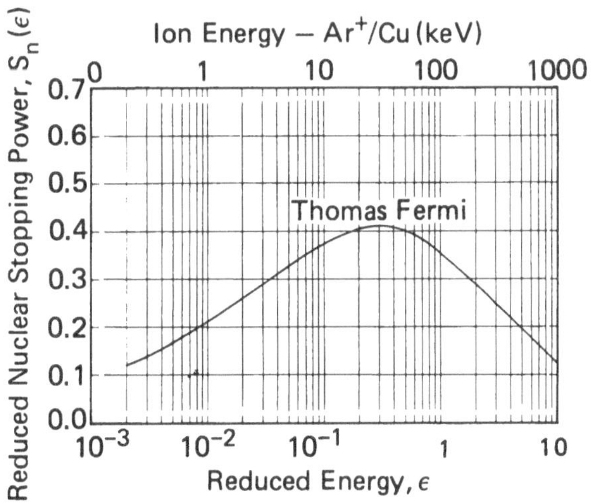

Fig. 21. Reduced nuclear stopping power, $s_n(\varepsilon)$, as a function of ε, bottom scale and of E for Ar^+—Cu, top scale. Based on Thomas-Fermi potential (see Equation 5). (Based on data presented in ref.[45])

the eV range. However, it has been extensively applied to polycrystalline targets over a wide energy range, and agreement with experiment to approximately a factor of two is often obtained. This agreement is quite impressive considering the fact that the theory contains no adjustable parameters in the usual sense. Moreover, much of the experimental work is subject to problems in interpretation because of dose effects, surface contamination, and surface roughness. Therefore, the real accuracy of the theory has not been determined to this author's satisfaction.

Equations 17 and 18 are quite inaccurate for light ions such as (H^+, D^+, and He^+); therefore, a few comments are in order. Weismann and Behrisch[87)] have demonstrated that backscattered ions make a large contribution to the sputtering yield for the irradiation of a heavy target with a light ion at high energy. They evaporated a thin film of copper onto substrates of Be, V, Nb, and Ta and subsequently measured the sputtering yield of the copper overlayer as a function of substrate material. The various materials produced different intensities of backscattered ions; therefore, the contributions to the copper sputtering yield from the incident and backscattered ions could be separated. They concluded that up to 50% of the sputtering events in bulk copper could result from backscattering, i.e., from ions going deep into the target and then being reflected back toward the surface.

Since a large fraction of the incident ions are reflected back through the surface (see Fig. 14), a theory for sputtering by light ions should consider: (1) electronic stopping, (2) large angle scattering, and (3) a surface correction. When processes 1 and 2 (see Ref. 54) are included in the theory, α (see Equation 16), contrary to the case for keV ions, depends sensitively upon the ion energy, i.e., the sputtering yield is no longer directly proportional to the nuclear stopping power. Comparison of experiment with the theoretical predictions[54)] shows the measured values to be approximately one-half of the calculated ones. The general shape of the yield curves, however, agree well with Weismann's predictions.

There is no well developed theory which describes the sputtering of multicomponent systems (e.g., oxides), but some authors (see, for example, Ref. 88) have assumed that the results of Sigmund's theory are valid and have applied reasonable values for the stopping power and binding energy to obtain results for various multicomponent systems.

2.2.4.2 Sputtering — Experimental

Important factors which affect the sputtering yield include: the surface structure, the mass ratio (M_1/M_2), the angle of incidence, and the energy. Table 2 and Figs. 22–24 are an attempt to present representative data indicating the influence of these parameters. As shown in Fig. 22, the sputtering yield rises rapidly from an apparent threshold which is usually somewhere between 10–30 eV (see also Ref. 90). From approximately 100 eV upwards, it becomes a slowly varying function of energy until it reaches a broad maximum and then it slowly decreases. The maximum (except for light ions) usually occurs at energies which are higher than would normally be encountered in a glow discharge. The sputtering theory of Sigmund, which was described in the preceding section, reproduces these trends quite well.

The angular dependence of the sputtering yield is qualitively similar for most materials (see Ref. 96). It increases then reaches a maximum and finally decreases as

Table 2. Sputtering yields for 28 elements under Ne^+ and Ar^+ ion bombardment. (From Ref. [89])

Target	Neon				Yield at lowest ion energy		Argon			
	100 (eV)	200 (eV)	300 (eV)	600 (eV)	Y	E (eV)	100 (eV)	200 (eV)	300 (eV)	600 (eV)
Be	0.012	0.10	0.26	0.56	0.05	80	0.074	0.18	0.29	0.80
Al	0.031	0.24	0.43	0.83	0.11	100	0.11	0.35	0.65	1.24
Si	0.034	0.13	0.25	0.54	0.06	80	0.07	0.18	0.31	0.53
Ti	0.08	0.22	0.30	0.45	0.081	100	0.081	0.22	0.33	0.58
V	0.06	0.17	0.36	0.55	0.03	60	0.11	0.31	0.41	0.70
Cr	0.18	0.49	0.73	1.05	0.026	40	0.30	0.67	0.87	1.30
Fe	0.18	0.38	0.62	0.97	0.064	60	0.20	0.53	0.76	1.26
Co	0.084	0.41	0.64	0.99	0.048	60	0.15	0.57	0.81	1.36
Ni	0.22	0.46	0.65	1.34	0.067	60	0.28	0.66	0.95	1.52
Cu	0.26	0.84	1.20	2.00	0.10	60	0.48	1.10	1.59	2.30
Ge	0.12	0.32	0.48	0.82	0.017	30	0.22	0.50	0.74	1.22
Zr	0.054	0.17	0.27	0.42	0.027	60	0.12	0.28	0.41	0.75
Nb	0.051	0.16	0.23	0.42	0.017	60	0.068	0.25	0.40	0.65
Mo	0.10	0.24	0.34	0.54	0.027	60	0.13	0.40	0.58	0.93
Ru	0.078	0.26	0.38	0.67	0.012	60	0.14	0.41	0.68	1.30
Rh	0.081	0.36	0.52	0.77	0.19	100	0.19	0.55	0.86	1.46
Pd	0.14	0.59	0.82	1.32	0.033	50	0.42	1.00	1.41	2.39
Ag	0.27	1.00	1.30	1.98	0.22	60	0.63	1.58	2.20	3.40
Hf	0.057	0.15	0.22	0.39	0.004	40	0.16	0.35	0.48	0.83
Ta	0.056	0.13	0.18	0.30	0.01	60	0.10	0.28	0.41	0.62
W	0.038	0.13	0.18	0.32	0.008	60	0.068	0.29	0.40	0.62
Re	0.04	0.15	0.24	0.42	0.034	80	0.10	0.37	0.56	0.91
Os	0.069	0.16	0.24	0.41	0.57	100	0.057	0.36	0.56	0.95
Ir	0.069	0.21	0.30	0.46	0.019	60	0.12	0.43	0.70	1.17
Pt	0.12	0.31	0.44	0.70	0.032	60	0.20	0.63	0.95	1.56
Au	0.20	0.56	0.84	1.18	0.035	50	0.32	1.07	1.65	2.43(500)
Th	0.28	0.11	0.17	0.36	0.017	60	0.097	0.27	0.42	0.66
U	0.063	0.20	0.30	0.52	0.14	100	0.14	0.35	0.59	0.97

one goes from normal incidence to grazing incidence. The decrease at grazing incidence is a consequence of the fact that many of the impinging ions are reflected away from the surface without transferring a significant fraction of their energy to the lattice. In a glow discharge environment, the plasma sheath usually tends to follow the contour of the surface and therefore most ions approach the surface at normal incidence.

2.2.4.3 Comments on Sputtering in a Glow Discharge

Sputtering in a glow discharge is an extremely important process since many surfaces exposed to the plasma are bombarded with energetic ions or neutrals. This state of affairs can produce positive or deleterious results depending upon the situation. For example, the deposition of films by glow-discharge sputtering is a widely employed technique which is very useful (see Ref. 97). In contrast, the deposition of previously sputtered impurities on surfaces which are also being sputtered or etched often

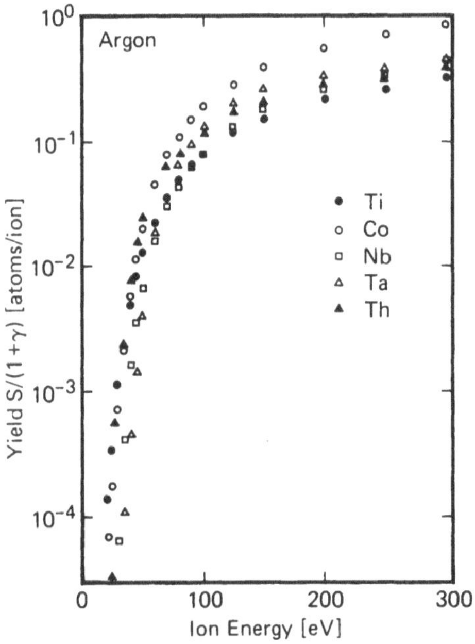

Fig. 22. Sputtering yield curves for argon. (From Ref.[90])

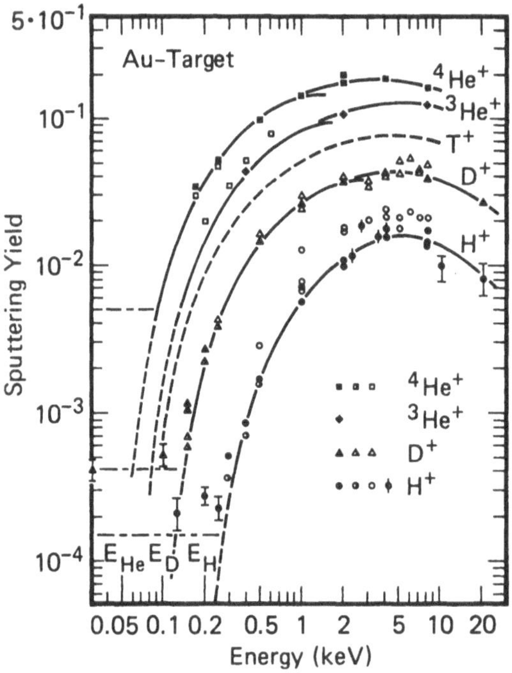

Fig. 23. Sputtering yield of gold for H^+, D^+, $^3He^+$, and $^4He^+$ ion irradiation as a function of energy. The data is from Refs.[91–94]. The original figure was published in Ref. 91 and the reader is referred to that paper for more detail. Theoretical threshold energies are indicated on the energy scale

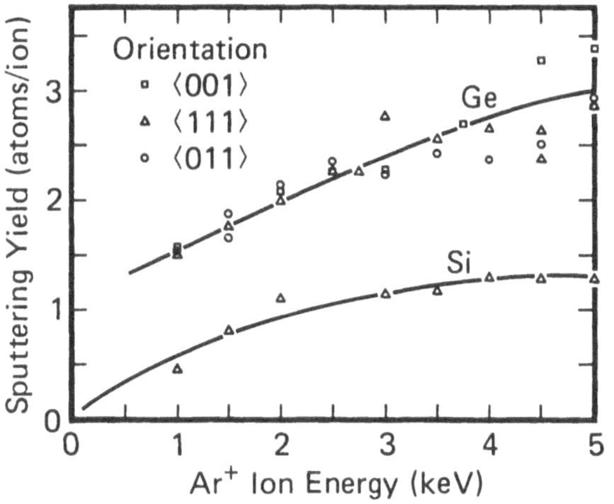

Fig. 24. The sputtering yields of Si and Ge as a function of Ar^+ ion energy. Normal incidence. (From Ref.[95])

produces undesirable results[98]. Moreover, unexpected phenomena often occur. For example, negative ions sputtered from the target (see Fig. 1) are accelerated across the discharge and collide with the anode. The sputtering produced by these negative ions not only reduces the amount of target material deposited on the anode, but, in some instances, produced steady state sputter removal of the anode material itself[99-101]. A wide variety of sputtering-induced phenomena occur in plasma environments, and it is not possible to discuss all of them within the confines of the space which is available. However, investigators should always interpret their plasma data with an awareness that sputtering has occurred within the system.

Sputter yields as well as trapping and reflection coefficients must be used with a certain degree of caution when interpreting glow discharge data because the appropriate parameters are usually not well defined. For example, charge exchange reactions in the sheath affect the energy of the ions which collide with the surface[102]. Since charge exchange reactions often have large cross sections, this type of process is quite probable. Moreover, the charge exchange reaction producès a flux of energetic neutrals which also collide with the surface. Therefore, the decreased sputtering rate resulting from a decreased ion energy which is a consequence of the charge exchange reaction tends to be compensated by an increased sputtering rate due to the flux of energetic neutrals. Neither the theoretical nor the experimental description of this situation is advanced enough to allow a quantitative description.

The maximum energy of ions bombarding a given surface is equal to the difference between the plasma potential and the surface potential. The plasma potential is seldom measured and is often different than one would intuitively expect. (For a discussion of plasma potentials in an rf discharge see Ref. 103. For measurements of plasma potentials in a number of commercial systems see Ref. 104.) For example, it is possible for a system of the type shown in Fig. 1 to have the average dc potential of both the anode and target close to ground. The plasma potential is then several hundred volts positive. (Some commercial systems operate in this mode[104].) In this situation, all surfaces exposed to the discharge are bombarded with positive ions and hence sputtered. Since sputtered material usually condenses on the surrounding surfaces, this can cause severe impurity problems in plasma-deposited-films or in films deposited by glow discharge sputtering. Moreover, in plasma etching, one can envision a situation where non-etchable material is deposited upon a surface being etched, thus causing severe surface roughness and eventually causing the etching reaction to cease. In addition, impurities deposited upon a surface which is being sputtered can cause the formation of conical structures which in turn makes the surface very rough[98,105,106]. This can decrease the sputtering yield[107].

Another factor which influences sputtering yields is the inevitable presence of impurity gases in the discharge. These impurity gases tend to reduce the sputtering yields below the clean surface values reported above[108-110]. This happens because impurity gases, which are being chemisorbed on the surface, tend to inhibit the sputtering of the underlying substrate.

2.2.5 The Altered Layer

When a solid containing more than one element is bombarded with ions, the constituents are initially sputtered at different rates producing a thin layer of altered chemical composition at the surface[111-118]. For example, many saturated oxides are reduced to a lower oxide or to the metal itself by ion bombardment[111,112]. Moreover, the energy input from the ions tends to allow the surface region to relax toward a state of thermodynamic equilibrium. This tendency is illustrated by the $CoFe_2O_4$ system[111] where bombardment with Ar^+ ions produces a deficiency of oxygen and reduces both metals. However, cobalt is reduced much more than iron because its oxide is less stable. However, the tendency toward equilibrium never proceeds to completion because the lattice is also disrupted by high energy collisions. For example, the widely observed presence of noble gases in the surface region of an ion bombarded sample is good evidence for an thermodynamically unstable state. Therefore, it must be concluded that surfaces exposed to ion bombardment in a glow discharge generally have undefined chemical properties.

Surfaces held at large negative potentials with respect to the plasma (such as the target in Fig. 1) are bombarded with ions whose energy varies from a few hundred to a few thousand eV. Grounded and floating surfaces are often bombarded with ions whose energy is some tens of eV. In some instances, the geometry of the walls and/or of the mechanical fixtures exposed to the discharge cause the plasma potential to stay well above ground and in those instances even grounded surfaces will be bombarded with high energy ions. The change in chemical composition of the surface due to this ion (or electron) bombardment will influence various surface properties such as its chemical reactivity toward gas phase particles and the secondary-electron coefficient.

The relationship between the surface concentration and bulk concentration and also the depth of the altered layer will be the topics discussed in the following paragraphs. The amount of energy transferred from the impinging ion to a target atom is small. Even for particles in the MeV range, the average energy transferred to a target atom is only a few hundred eV[119]. Kornelsen's data[65] indicates that an atom with this small energy will not travel far through a lattice. Therefore, sputtered atoms should originate predominantly from the top surface layer. This expectation is consistent with the sputtering model of Sigmund[35] and also with the simulation studies of

Harrison et al.[120]. Harrison's results indicate, for example, that sputtered atoms originate primarily from the top atomic layer even at ion energies up to 10 keV.

Since sputtered atoms originate primarily from the top atomic layer, it is reasonable to assume that the sputtering rate of a given constituent is proportional to its surface concentration. This leads to the following equation:

$$R_1 = \nu^+ \Gamma_1 \theta_1 , \tag{20}$$

where ν^+ is the ion flux, Γ_1 is the sputtering yield, θ_1 is the surface concentration, and R_1 is the rate that atoms of a given type are being removed from the surface. The subscript designates the component.

Under steady state conditions, the relative removal rates of material from the surface must reflect the bulk concentration. This criteria is the driving force which determines the concentrations of the various constituents at the surface. For a two component system, this implies that $C_1/C_2 = \Gamma_1\theta_1/\Gamma_2\theta_2$ where C_1 and C_2 are concentrations of the two constituents in the bulk material. Therefore,

$$\frac{\theta_1}{\theta_2} = \frac{\Gamma_2}{\Gamma_1}\frac{C_1}{C_2} . \tag{21}$$

When a steady state condition has been achieved, Equation 21 implies that the relative surface concentrations are only functions of the bulk concentrations and the sputtering coefficients. This point cannot be overemphasized. Many authors have misinterpreted their data because they did not understand the consequences of this result. Once the sputtering coefficients are known, then thermodynamic properties, such as a tendency towards surface segregation, do not affect the surface concentration. However, the sputtering yields themselves are partially determined by binding energies and the type of compounds which are present in the surface region. These parameters are, of course, influenced by thermodynamic considerations.

During irradiation with ions, atoms are continually removed from the surface by sputtering leaving surface vacancies. There are also a large number of vacancies and interstitials which migrate to the surface from the bulk. As a consequence, the surface region is in a highly mobile state. Under these circumstances, a gradient in chemical potential will influence the flow of matter toward the surface. This phenomenon along with cascade mixing is, in our opinion, the most probable mechanism by which the elemental composition of the second, third, and subsequent layers can be altered by ion bombardment. The thickness of the altered layer would then be determined by the depth of the surface region where mobility is significant. The effects of cascade mixing on the surface region of a solid bombarded by ions has recently been analyzed in an excellent paper by Andersen[120a].

The depth of the altered layer, in our opinion, can be estimated by the following procedure. The energy (momentum) pulse is likely to penetrate the lattice more deeply than does the particle itself. For example, 200 eV Xe^+ ions collide with a tungsten lattice and are primarily reflected back into the gas phase. Nevertheless, atomic motion induced by the impinging ion probably occurs to a depth of 30–50 Å beneath the surface. Therefore, a crude estimate (but probably the best available) of the depth of the altered layer can be obtained by assuming a constant value of 30 Å at energies where ions do not significantly penetrate the lattice. The depth of the altered layer then should increase with energy at a rate proportional to the projected range.

An estimate for the projected range ($\bar{R}_p$) is obtained in the following manner. As a rule of thumb, the equation $\bar{R}/\bar{R}_p = 1 + (^1/_3)\ M_2/M_1$ is a fair approximation at low energies[121]. $\bar{R}$ is the range which can be obtained from the data shown in Fig. 25.

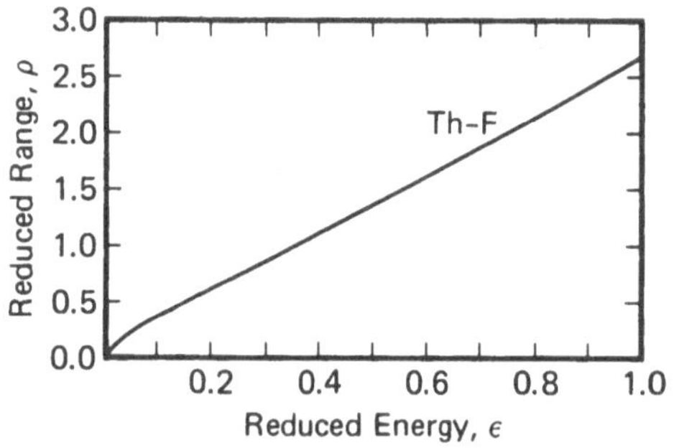

Fig. 25. Reduced range (ϱ) vs. reduced energy ε. (From Ref.[33])

$$\varrho = \bar{R} \cdot NM_2 4\pi a^2 \left[\frac{M_1}{(M_1 + M_2)^2}\right]; \qquad \varepsilon = \frac{aM_2E}{Z_1Z_2e^2(M_1 + M_2)}$$

$\bar{R}$ is the range and E the incident ion energy. See Sect. 2.2.2 for definitions of symbols. This curve is calculated assuming a Thomas-Fermi potential and with neglect of electronic stopping

That these ideas have some merit is indicated by the work of Hart, Dunlap, and Marsh[122]. These investigators deposited a fraction of a monolayer of copper onto a silicon wafer and then monitored the position and concentration of the copper using Rutherford backscattering. After deposition, the copper, which was then located on the immediate surface, was bombarded with 20 keV Ne^+ ions to a fluence sufficient to sputter 90 Å of Si from the surface. The Rutherford backscattering spectrum, which was taken after this bombardment, showed that the copper was uniformly distributed to a depth of 600 Å which corresponds roughly to the projected range of the Ne^+ ions, i.e., the depth of the altered layer was approximately equal to the projected range of the Ne^+.

The fact that the constituents of a multicomponent system are not removed stoichiometrically by sputtering not only influences the altered layer but has also been used to control the composition of sputtered films, which have been grown in a plasma environment. A quantitative model, which relates composition to sputtering coefficients, has been published[114].

When the system described in Fig. 1 is used in a sputtered-film-deposition-mode, material is sputtered from the target and deposited upon the anode. Material previously deposited upon the anode is resputtered when and if the anode is biased with respect to the plasma so as to increase the number and energy of ions which are colliding with the growing film. Materials with high sputtering coefficients are preferentially removed which in turn leads to a change in film composition. This technique has been used to keep impurity gases out of films[70, 123] and also affect the composition of alloys[124].

2.2.6 Ion-Induced Chemical Reactions

2.2.6.1 Cooperative Effects in Ion-Induced Chemical Reactions

Chemical reactions between adsorbed species and substrate material, which are induced by ion bombardment (e.g., Ar^+), will be discussed in this section. Consider, for example, the XeF_2—Si—Ar^+ system. The reaction probability for XeF_2(gas) with silicon at room temperature is significant[125]. However, simultaneous bombardment with argon ions greatly enhances the probability that the incident fluorine from the XeF_2 will react to form SiF_4(gas). Reactions involving this type of synergistic effect are the topic of this section.

Recent investigations in the field of plasma etching have almost universally found that surfaces subjected to ion bombardment (e.g., the target surface in Fig. 1) react (i.e., etch) much more rapidly than those which are held near plasma potential. Examples of this type have been reported by Hosokawa et al.[126] for the halocarbon etching of silicon and by Holland and Ojha[127, 128] for the oxygen plasma-etching of carbon. Several systems exhibiting this type of behavior have also been observed in our laboratory[129].

It is this authors opinion, that *ion and electron enhanced chemical reactions of the type described above are probably important in MOST situations.* Nevertheless, to our knowledge, neither a theoretical understanding nor well defined experiments have been reported on this subject. Therefore, several experiments which have been performed recently in this laboratory will be described and then possible explanations for the ion-enhanced, chemical reactions will be discussed.

A molecular beam of XeF_2(gas) and a beam of argon ions were directed at the center of a silicon film which had been deposited on a quartz crystal microbalance. The sensitivity of the microbalance was such that the removal of one monolayer of silicon could be detected. In these experiments, the reaction products [e.g., SiF_4(gas)] were detected using mass spectrometry; the surface concentrations were detected using Auger spectroscopy; and the rate that material was being removed from the surface was measured with the microbalance.

The etch rate of silicon, which is determined from microbalance measurements, is plotted versus time in Fig. 26. The silicon is first exposed to XeF_2(gas) [$t < 200$ sec] and the spontaneous etching reported in Ref. 125 is observed. Then, the argon ion beam and the XeF_2 molecular beam were used simultaneously ($200 < t < 640$ sec) resulting in a greatly enhanced etch rate. Finally, the argon ion beam was used by itself ($t > 640$ sec) and physical sputtering was observed. The co-operative interaction between the Ar^+ beam and the XeF_2 is shown clearly in Fig. 26, in that, the etching of Si obtained with both Ar^+ ions and neutral XeF_2 molecules simultaneously incident upon the surface is about eight times the sum of the etching rates of each species measured separately. The data in Fig. 26 and a knowledge that the sputtering yield of 450 eV Ar^+ is ~ 0.5 allows one to estimate that each incident argon ion produces as many as ~ 30 SiF_4 molecules per argon ion.

Similar conclusions can be drawn from the data shown in Fig. 27 where the partial pressure of SiF_4 is monitored mass spectrometrically as an Si sample is rotated into a jet of XeF_2 gas with and without ion bombardment. The ion-enhanced etching of silicon is indicated by the much larger evolution of SiF_4 when the sample is exposed

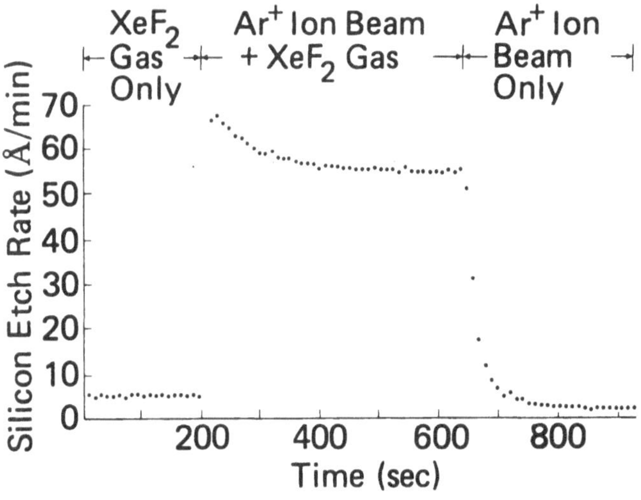

Fig. 26. Ion-assisted gas-surface chemistry using Ar^+ and XeF_2 on silicon (Volatile reaction product.).
Ar^+ energy = 450 eV
Ar^+ current = 0 (t < 200 sec)
= 2.5 μa (t > 200 sec)
XeF_2 flow = 2×10^{15} mol/sec (t < 660 sec)
= 0 (t > 600 sec)
(The Ar^+ current density and the XeF_2 flux are not uniform over the Si surface. The effective area for the Ar^+ current and the XeF_2 flux are estimated at 0.1 cm^2 and 0.3 cm^2, respectively.) (From Ref.[129])

to XeF_2 with argon ion bombardment as compared to XeF_2 exposure alone. Similar ion enhanced chemistry has been seen in our laboratory for F_2 on Si, F_2 on C, Cl_2 on Si, and O_2 on C. Moreover, the effect of radiation damage on the errosion of graphite by atomic hydrogen may be a related phenomena[130]. Furthermore, it will be shown in a subsequent section that electron irradiation produces similar effects.

We now direct our attention toward the understanding of possible mechanisms which produce ion-assisted etching. Whereas the thrust of this section is directed toward etching reactions, it should be emphasized that many of the concepts are also applicable to chemical reactions which produce involtaile products.

As was suggested in a previous paper[131], the steady state etching of solid material by exposure to gas phase particles with or without a plasma is usually described by the following sequence of steps: (1) nondissociative adsorption of gas phase species at the surface of the solid being etched; (2) dissociation of this absorbed gas (i.e., dissociative chemisorption); (3) reaction between adsorbed radicals and the solid surface to form an adsorbed product molecule, e.g., SiF_4(ads); (4) desorption of the product molecule into the gas phase; and (5) the removal of nonreactive residue (e.g., carbon) from the surface.

The first step always occurs; since, attractive forces between the undissociated molecule and the surface usually exist. This step may involve adsorption into a so-called "precursor state" where the molecule is mobile and diffuses across the

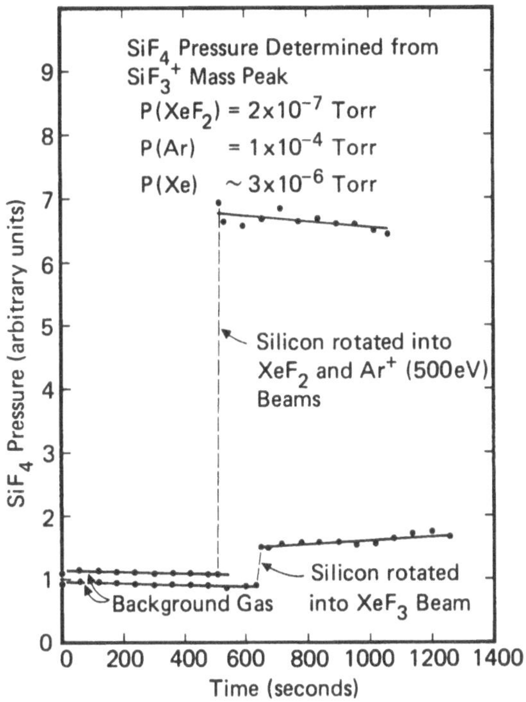

Fig. 27. The SiF_4 pressure as a silicon surface is rotation into a beam of XeF_2 molecules with an without Ar^+ ion bombardment. The SiF_4 pressure was determined mass spectrometrically. The ratios of the pressure increases (top curve to bottom curve) indicate the magnitude of ion-assisted etching. The experiments were performed in a standard UHV system of the type often used for surface experiments. The experimental arrangement was similar to that described in Ref.[129])

surface until it dissociates; possibly at a step, kink, vacancy, or other defect. The first two steps would be combined and labeled "adsorption" for atoms but the conceptual framework would be the same.

The relationship between steps 2–5 and the mechanisms which cause ion-enhanced etching will now be discussed. There are at least three processes whereby ion bombardment may increase the rate of dissociative chemisorption. First, dissociative chemisorption may occur exclusively or preferentially at defect sites produced by the impinging ion. Therefore, in instances where dissociative chemisorption is the rate limiting step, ion bombardment will significantly enhance the overall reaction rate. We suspect that phenomena of this type occur in a wide variety of plasma situations; and yet, quite frequently remain unnoticed. Secondly, ion and/or electron bombardment may cause an adsorbed molecule to dissociate in situations where this would not normally happen. The reactive freagments, which result from this collision, often produce product molecules. Thirdly, dissociative chemisorption often occurs in a clean surface, whereas it may not occur on the same surface when it is covered by a monolayer of adsorbed gas. Removal of the adsorbed layer by electron bombardment (electron-stimulated desorption) or ion bombardment (physical or chemical sputtering) may allow the dissociative chemisorption reaction to proceed, which in turn promotes etching. For example, it is expected that a monolayer of oxygen, carbon, or chlorine would slow the etching of silicon exposed to a flux of fluorine atoms. When the adsorbed layer is removed by radiation, then the etch rate should be enhanced.

If a surface reaction is to involve more than monolayer-chemisorption, then the species adsorbed on the surface must be able to migrate into the second and deeper layers forming new chemical bonds and often new molecular species. This is step 3, product formation, and it often requires an activation mechanism to proceed, i.e., a monolayer is formed and the reaction stops unless the substrate is held at elevated temperature or there is ion or electron bombardment. Damage-enhanced diffsusion, knock-on collisions, and bond breaking may promote the reaction in the presence of ion bombardment. Although the precise mechanisms are unclear, it is certain that electron and ion bombardment cause step 3 to occur in some instances where the chemical reaction does not proceed in the absence of radiation.

The consequences of the above processes are largely determined by the volatility of the species which are formed. If the species are volatile, they will desorb causing etching of the surface (step 4). If the species are involatile, a layer of reaction product will form with its thickness being determined by the mobility of the chemically active species through the layer (e.g., the oxidation of aluminium). There are several mechanisms which could cause radiation-enhanced etching by influencing step 4. For example, an involatile product molecule may be weakly bound to the surface and as a consequence of this weak binding the sputtering yield of the product moecule could be substantially larger than the sputtering yield of the original substrate. Therefore, a combination of physical sputtering and chemical reaction would produce a radiation-enhanced etch rate which is substantially larger than would be expected from consideration of the separate individual etch rates[132]. Moreover, for involatile products (e.g., oxides), the ion bombardment may cause radiation-enhanced diffusion which allows the oxygen to diffuse into the lattice thus causing an increase in the reaction rate at the surface.

Finally, one must consider the possibility of an involatile residue being formed by one or more of the atomic species present in the molecular gas, e.g., adsorbed carbon from CF_4. Unless this residue is removed by some mechanism (step 5), it will terminate the etching reaction. Electron and ion bombardment could enhance residue removal by many of the same mechanisms mentioned above in connection with steps 2–4.

2.2.6.2 Chemical Reactions Involving Constituents of the Incident Ion

When a molecular ion collides with a solid surface, it is usually dissociated into its various constituent atoms. Some of these atoms (or radicals) are reflected away from the target and others are incorporated into the lattice. The reflected atoms are likely to react with the surrounding surfaces and may be incorporated into growing films. That an ion will dissociate upon impact with a surface is quite clear when it is realized that the kinetic energy of the ion is often many times greater than the energy required to break a chemical bond. Nevertheless, there have only been two investigations of the probability that an ion will be dissociated as a function of its kinetic energy[133, 134]. Both of these have involved nitrogen.

Figure 28 shows that the probability for dissociation of an N_2^+ ion when it collides with a surface has a threshold around 9 eV, which is about the energy required to break the bond between the two nitrogen atoms. The probability reaches unity for energies greater than ~ 100 eV. It was also shown in this set of experiments that the

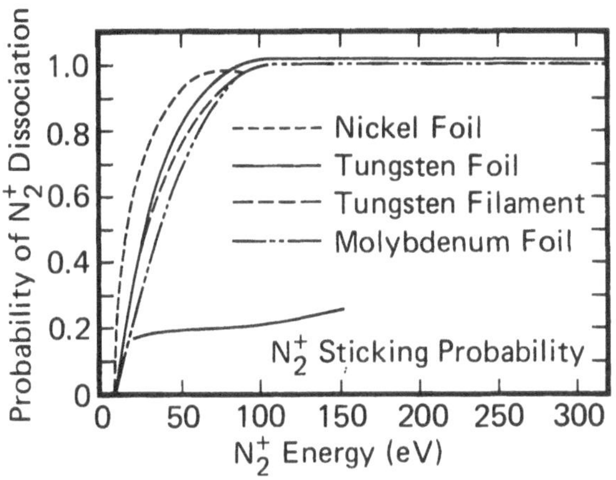

Fig. 28. Probability that an N_2^+ ion will dissociate upon collision with a surface versus ion energy. The probability of dissociation is equivalent to the probability that the N_2^+ will react chemically as a consequence of its collision. When is reacts chemically, it disappears from the gas phase.

probability that nitrogen atoms from N_2^+ would remain trapped in the target surface (i.e., the sticking probability) was $\sim$0.2. Therefore, about 80% of the atoms from the N_2^+ ion were reflected back into the gas phase and subsequently reacted with the surrounding surfaces. About 20% of the atoms remained at the target surface where they formed a thin nitride layer. Figure 29 shows the number of atoms in the nitride layer as a function of the number of bombarding ions[135]. The sticking probability is obtained from the initial slope.

Ion bombardment of surfaces can also produce significant quantities of chemically reactive atoms and radicals by other mechanisms. For example, consider the glow discharge system in Fig. 1 where the target is tungsten, the argon pressure is 10^{-1} Torr, and the partial pressure of nitrogen is 5×10^{-6} Torr. Molecular nitrogen is dissociatively chemisorbed on the surface and then is subsequently sputtered in the form of

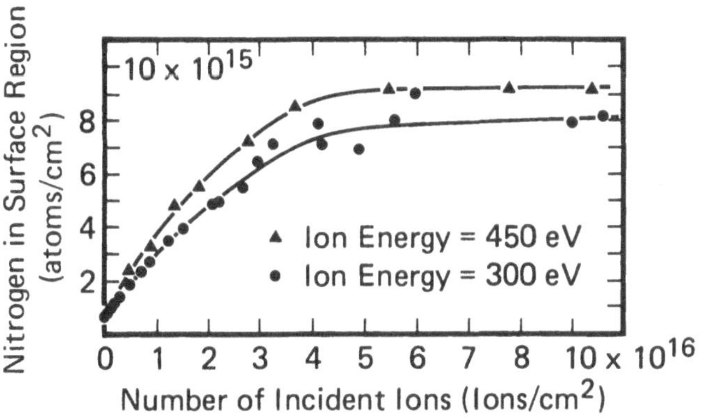

Fig. 29. Amount of nitrogen contained in the surface region of tungsten which was bombarded with N_2^+ vs. ion dose. It was estimated that a nitride layer of approximately 30–100Å was formed at the surface. (From ref.[135])

nitrogen atoms by the impinging argon ions. Hence the tungsten surface is a continual and probably dominant source of neutral nitrogen atoms.

In summary, one must expect that surfaces in a plasma which are exposed to bombardment with chemically reactive ions (e.g., N_2^+, O_2^+, CH_4^+, etc.) will have a surface region of modified chemical composition whose thickness is at least 20–30 Å but which is not less than the projected range of the ions. For example, bombardment of a metal surface with O_2^+, N_2^+, or CH_4^+ will probably produce a surface layer of oxide, nitride, and carbide, respectively. Moreover, in some situations ion bombardment (for example with CH_4^+ and CF_3^+) produces a surface layer of arbitrary thickness which contains primarily carbon[136]. Results of this type are likely to occur when one or more constituents of the ion are non-volatile and the ion-energy is relatively small.

Finally, it should be remembered that a surface bombarded by ions will be a source of neutral atoms and molecular radicals which can react chemically with the surrounding surfaces.

3 The Interaction of Electrons with Surfaces

The interaction of an electron with a surface produces at least three phenomena which are important in a plasma environment. They are: (1) chemical reactions between gas phase species and a surface where electron bombardment is required to activate the process, (2) electron-induced secondary-electron emission, and (3) electron-induced dissociation of sorbed molecules. A fourth phenomenon — lattice damage produced by energetic electrons[137–139] — depends sensitively upon the properties of the material being bombarded, and, it is important in specialized situations, but it will not be discussed in this paper.

The influence of secondary-electron emission on phenomena related to a glow discharge can be illustrated by considering the system described in Fig. 1. Ions arriving at the target produce secondary electrons which are accelerated across the plasma and collide with the anode. An electron, which collides with the anode, generates a flux of energetic secondary electrons which may produce radiation damage in the lattice and/or induce chemical reactions in the surface region. Moreover, some of the secondaries are emitted into the gas phase where they are accelerated across a thin sheath into the plasma and subsequently produce ionization and excitation.

3.1 Secondary-Electron Emission

The treatment of secondary-electron emission presented in this section essentially follows the ideas and organization of Sickafus[140] which in turn was strongly influenced by Wolff[141]. Upon entering a solid, an energetic electron is subjected to a sequence of elastic and inelastic scattering events. The collisions produce a cascade of moving electrons. The intersection of the cascade with the surface results in the emission of electrons into the gas phase. Therefore, the externally-observed, energy-

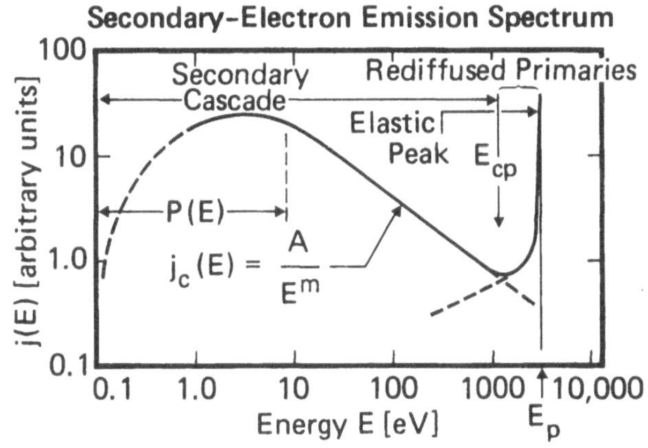

Fig. 30. Schematic representation of the backscattered secondary-electron spectrum associated with an (isolated) external source of monoenergetic electrons of energy E_p. This log j(E) vs. log (E) display mode emphasizes the separation of the spectrum into three parts: (1) the secondary cascade which is bounded at high energies (at E_{cp}) by (2) the region of rediffused primaries which is bounded by (3) the elastic peak. At low energies the cascade is attenuated by the escape probability P(E). Auger and characteristic loss processes, among other things are not included in this idealized spectrum. (From Ref.[140)])

dependent emission current j(E), can be visualized as a sampling of an internal distribution, N(x, E′).

Figure 30 schematically represents a typical backscattered secondary-electron spectrum, j(E), associated with an *external* source of monoenergetic electrons of energy E_p. The spectrum is divided into three sections: the secondary-electron cascade, rediffused primaries, and the elastic peaks. At the low energy end of the cascade, a pronounced deviation from linearity occurs because the escape probability P(E′) becomes small. This evolves into a linear region at higher energies where the escape probability is almost unity. Another deviation from linearity occurs at high energy where reflected primary electrons, which have lost some energy, become the dominant source (the region of rediffused primaries). Finally, there is a very intense peak of primary electrons which are reflected back into the gas phase with all of their initial kinetic energy.

Wolff[141)] and more recently other investigators[142, 143)] have applied the Boltzmann diffusion equation to a description of the secondary-electron cascade. This approach is quite satisfying because it has a clearly defined foundation which seems to encompass all of the basic physical processes needed to describe the situation. It also yields an approximate solution in analytic form which is given by Equation 22.

$$j(E) = j(E')\,P(E') \propto \frac{P(E')}{n_c\sigma_0(E')\left(1 - \frac{7}{5}\frac{E_F}{E'}\right)}\left(\frac{E_P}{E'}\right)^x. \qquad (22)$$

A proportionality symbol has been used in Equation 22 because the original equation (see Ref.[141)]) seems to have a problem with units. In this equation, $\sigma_0(E')$ is the inelastic electron-electron scattering cross section, E_F is the Fermi energy, E_P the energy of the incident electrons and n_c the number density of conduction band electrons. $E' = E + W$ where W is the inner potential and E is the energy of the emitted electron in the gas phase. The exponent x is energy dependent below $E' \sim 4E_F$ and is essentially constant for greater energies:

$$X \cong 2\,, \qquad E' \geqq 4E_F\,.$$

The external current leaving the surface, j(E), is related to the internal current approaching the surface, j(E'), by the escape probability, P(E'). According to Wolff, $P(E') = 1 - [(\phi + E_F)/E']^{1/2}$ for a spherically symmetric distribution. ϕ is the work function and E_F the Fermi energy.

For energies $E' \gg E_F$, the factor P(E') in Equation 22 and the factor containing E_F make negligible contributions to j(E). Therefore, $j(E) \propto [\sigma_0(E')^x \sigma_0(E')\, E^x]^{-1}$. Assuming that $\sigma_0(E)$ is of the form, $E^{-1} \log E$, then

$$j(E) = \frac{CE_p^x}{[E^{x-1} \log E]} \tag{23}$$

where C is a constant.

Wolff suggests that the assumptions made in obtaining Equation 22 limit its applicability to energies less than 100 eV. However, Sickafus[140)] notes that Wolff's solution seems to have applicability to energies higher than originally anticipated.

Secondary-electron coefficients are strongly dependent upon the condition of the surface. The presence of adsorbed gas or surface roughness can significantly alter the number of secondary electrons. Moreover, much of the work in this field predates ultra-high-vacuum technology and the associated surface-characterization tools (for reviews see Refs. 144–146). In addition, surfaces exposed to a plasma are not well characterized. Therefore, crude, estimates of the magnitude of the secondary-electron coefficients seem to be the most useful type of data in the present context.

According to Hachenberg and Brauer[144)], the maximum yield for metals, δ_m, is usually between 0.6 and 1.7. The maximum yield for insulators and semiconductors cover a much wider range: $1 < \delta_m < 20$. At the lower limit of this range we have the well-known semiconductive elements Ge, Si, Se and also compound semiconductors such as Cu_2O and PbS. Substances with high yields include intermetallic compounds, alkali halides, alkali oxides, and alkali earth oxides.

3.2 Electron-Induced Dissociation of Sorbed Molecules

The bombardment of surfaces with electrons causes the dissociation of molecular surface complexes. Some of the fragments enter the gas phase and others remain attached to the solid. The first process is called "Electron-Stimulated-Desorption" (ESD) and has been widely investigated. There have been several recent reviews[147–149)]. The latter process has been widely recognized and often leads to deposition of films (e.g., carbon) upon the surface. Nevertheless, few if any well defined investigations of this phenomena have been reported.

Electron-stimulated out desorption occurs with high probability from a large variety of metals. For example, adsorbed layers of halogens, the alkali metals, hydrogen, oxygen, and carbon monoxide often have large cross sections. To our knowledge, ESD does not occur for most clean metals or for chemisorbed nitrogen. However, the surface of many oxides and hydroxides becomes deficient in oxygen under electron bombardment. Moreover, a hole can be drilled through an alkali halide lattice by directing an electron beam onto its surface.

Madey and Yates[147)] have summarized the major experimental observations related to electron-stimulated-desorption. The following section has been modified to reflect the present author's opinion but in general tends to closely follow the organization and conclusions of their original paper.

1) Bombardment of solid surfaces with electrons can cause desorption of ground-state neutrals (both atoms and molecules), ions, and metastable species. In addition, dissociation of adsorbed molecules with the resulting fragments remaining attached to the surface can be induced by electron bombardment. Conversion of one bonding mode to another can also occur.
2) The flux density of desorbed species is, typically, linearly proportional to the bombarding current density, indicating that desorption proceeds via a direct excitation mechanism.
3) The cross sections for ESD processes on most surfaces are usually much smaller than cross sections for comparable gas phase processes involving electron-induced dissociation and dissociative ionization[150)]. This may be a consequence of the fact that many fragments remain adsorbed on the surface and/or that non-radiative processes such as those described in Sect. 2.1.1 cause the molecule to de-excite before it dissociates. For 100 eV electrons, typical cross sections for gas-phase dissociation are 10^{-16} cm^2 (see Ref. 150). For most adsorbates, cross sections lie in range of 10^{-18} to 10^{-23} cm^2. A few examples of higher cross sections for adsorbed layers are known, and many examples of smaller cross sections exist.
4) Typical cross sections for desorption of ions are smaller (by a factor of ten or more) than cross sections for neutral desorption.
5) Cross sections for desorption of adsorbed species are very sensitive to the mode of bonding. In general, weakly bound adsorbates have a much higher cross section than adsorbates which are much more strongly bound.
6) The binding energies of even low mass adsorbates are sufficiently large so that direct momentum transfer does not cause desorption. This fact, coupled with the observation that ESD ions have large energies indicates that ESD proceeds via an electronic excitation mechanism. The energy dependence of the cross sections are quite consistent with this conclusion[149)].

Dissociation at a surface appears to be analogous to dissociation in the gas phase. The impinging electron causes a Franck-Condon transition to an electronic state which subsequently dissociates. This one-dimensional Franck-Condon excitation model is illustrated schematically in Fig. 31. The cross section for the electronic transition is probably comparable to gas phase excitation processes. After excitation the particle, which is now in a repulsive state, begins to move away from the surface. If it has sufficient energy it may escape from the surface. If not the fragments remain adsorbed. Moreover, radiationless de-excitation may occur

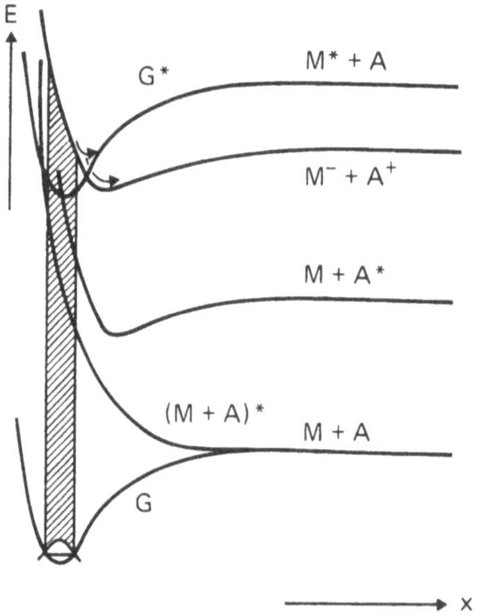

Fig. 31. Schematic potential energy diagram for interaction between absorbate A and a surface M. G is the ground state of the molecular complex, $M^- + A^+$ is an ionic state, $(M + A)^*$ is an antibonding state, $M + A^*$ is a state where the adsorbate is excited and the substrate is in its ground state, $M^* + A$ is a state where the substrate is excited and the adsorbate is in its ground state. Possible electronic transitions from the ground state G to the various excited states are indicated by the shaded Franck-Condon region. Electron bombardment can presumably excite any of these states. (From Ref.[147])

so rapidly that an electronic state which would dissociate in the gas phase does not lead to dissociation when the molecule is near the surface.

Excitation to the repulsive electronic state may also involve a multi-electron process. For example, creation of a core hole on a metal atom in an oxide may lead to an interatomic Auger transition which ultimately results in a positive oxygen ion which desorbs because it is now in a strongly repulsive Madelung well. Knotek and Feibelman[151] have reported results which they interpret in this manner. Core ionization in the adsorbed molecule can also lead to an Auger process which leads to desorption.[151 a]

There are also other mechanism which can lead to desorption. For example, the generation of electron hole pairs by either photons or electrons can produce desorption from some insulators and semiconductors. The holes are believed to reach the surface where they neutralize adsorbed negative ions which are subsequently desorbed into the gas phase. It is also possible that other defects (such as migrating H centers in the alkali halides) may cause desorption when they reach the surface[152]. Moreover, interstitial atoms generated within the solid may diffuse to the surface where they are desorbed[153].

Electron-induced desorption produces many adverse effects in a discharge system. When a glow discharge is ignited, large quantities of impurities are always emitted into the gas phase. In our opinion, this is primarily a consequence of electron-induced and ion-induced desorption. Whether ions or electrons dominate depends upon the situation; nevertheless, both processes make important contributions. The impurity gas concentration in the discharge decreases with time but always remains significant. It has been our experience that an argon glow discharge in a clean, metal-gasketed vacuum system usually has an impurity concentration of between 0.1–1 at.%. Under the very best conditions, it appears that at least one equivalent monolayer of impurity gas (10^{15} particles/cm^2-sec) is bombarding the surrounding surfaces each second. The impurity problem is much more severe when chemically active gases such as hydrogen, oxygen, or the halogens are used in the discharge. If a discharge experiment related to surface reactions is sensitive to impurities then interpretation of the results should be made with caution.

Electron beam induced dissociation of adsorbed species also yields fragments which may remain adsorbed on the surface. The consequences of this reaction are varied and depend strongly upon the situation. For illustrative purposes, two different situations are mentioned. Dissociation of weakly adsorbed molecules such as CCl_4, CF_2Cl_2, CF_3Cl, and CH_4 by electrons bombarding a silicon surface often produce a strongly-bound, carbonaceous surface layer[131]. This type of behavior is expected to occur on a wide variety of surfaces. In contrast, exposure of silicon to an electron beam in the presence of oxygen causes O_2 to be dissociated and also promotes the diffusion of the resulting atomic oxygen into the bulk material[153]. A surface oxide is formed.

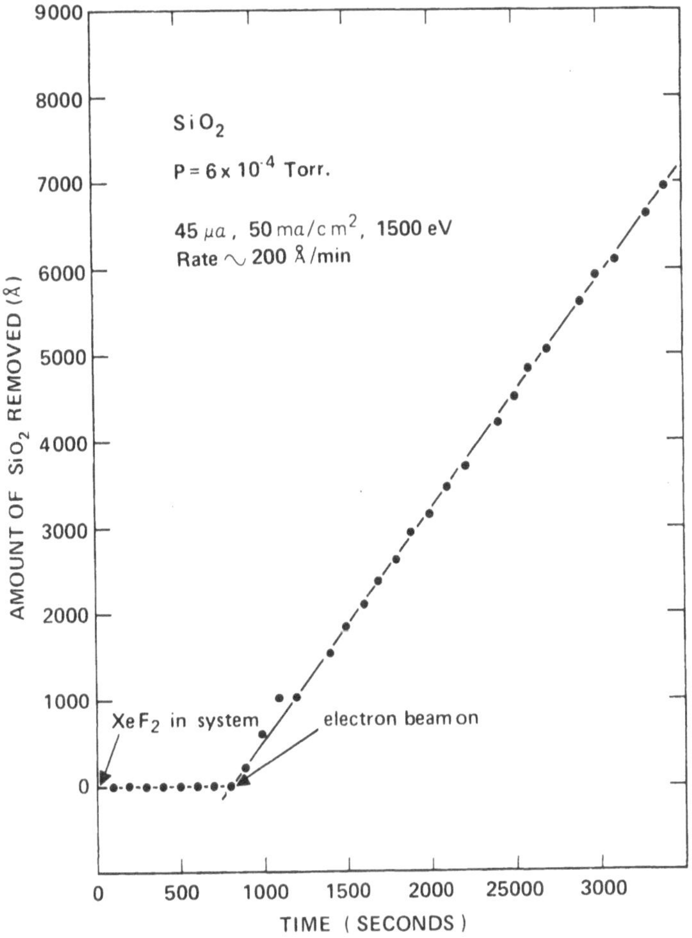

Fig. 32. Electron-assisted gas-surface chemistry using 1500 eV electrons and XeF_2 simultaneously incident on SiO_2. P(total) = 6×10^{-4} Torr with most of the ambient gas being xenon. Neither exposure to XeF_2 nor an electron beam produced etching by itself. Simultaneous exposure produces an etch rate of ~200 Å/min. (See Ref.[129])

3.3 Electron-Induced Chemical Reactions

It is the suspicion of the present author that gas-solid chemical reactions, which happen only in the presence of ion or electron bombardment, are a *widely* occurring phenomena in a plasma environment. Nevertheless, there are few if any well defined investigations of this topic. This is a consequence of the fact that most reactions of this type (e.g., oxidation or nitridization) produce a thin layer of non-volatile reaction product on the surface and then the reaction stops. A chemical reaction of this type is very difficult to investigate space experimentally. However, when the product is volatile, then the reaction is much easier to study since it continues indefinitely. Therefore, electron-induced chemical reactions will be discussed using an example where the reaction product is volatile.

Exposure of SiO_2, Si_3N_4, or SiC to XeF_2(gas) produces an adsorbed layer of fluorine[125]. The xenon does not remain on the surface, but it is immediately desorbed into the gas phase. This is all that happens in the absence of electron bombardment. However, in the presence of electron bombardment, SiF_4(gas) and other volatile products are produced[129]. Since the reaction products are removed from the surface, the reaction proceeds until all the material is volatilized. This is illustrated for the case of SiO_2 in Fig. 32. Similar data has been obtained for Si_3N_4 which reacts faster and for SiC which reacts slower. These are examples of a class of chemical reactions which require both active species (in this case fluorine) and energetic radiation (in this case electrons).

Many of the mechanisms discussed in Sect. 2.2.6.1 with regard to ions may also apply to chemical reactions enhanced by electron bombardment. A discussion of that type will not be repeated in this section. However, a mechanism for the electron-enhanced etching of SiO_2 can be suggested on the basis of processes which are known to occur. It is known, for example, that electron bombardment of SiO_2 causes oxygen to be desorbed into the gas phase, i.e., electron stimulated desorption occurs[154,155]. The silicon which remains upon the surface can now be attacked by the XeF_2(gas) to produce SiF_4(gas). In this manner both oxygen and silicon are removed from the SiO_2 lattice and the material is etched. The chemistry involved is likely to be more complex, but this simple model illustrates a possible mechanism.

Whereas a detailed understanding of chemical reactions enhanced by electron bombardment is not currently available, experimenters utilizing a plasma should be aware of the fact that chemical reactions of this type are likely to occur.

4 The Interaction of Neutral Species with Surfaces

4.1 General Comments

In order to understand the details of the interaction of a plasma with a surface, one must understand how each individual species in the plasma reacts with the surface. The important questions are: Does a particle react chemically? If so, what is the reaction probability? How does the reaction probability vary with coverage and temperature? Do the chemisorbed species recombine to form stable molecules? If so, how efficiently? The answer to these questions is important for all species but it is most important for ground state molecules since their concentration is generally large. Unfortunately, there is very little theoretical knowledge about this subject. We are not aware of any gas-surface system where these questions can be answered theoretically. Even a qualitative understanding of the important parameters is usually absent. The use of simple concepts such as the Langmuir isotherm to model a plasma situation is generally not a useful exercise because the surface reaction is much more complicated than this or other simple isotherms would imply. The most useful approach at this time is to *experimentally* investigate the interaction of a known particle with a well defined surface and then to relate this information to processes which occur in a plasma. For ground state molecules, experimental information as to whether or not a given particle will chemisorb and estimates for sticking probabilities are widely scattered throughout the literature. Table 3 is a summary of the older literature which was originally compiled by G. C. Bond[156)].

The complexities associated with understanding sticking probabilities are aptly illustrated by the N_2 tungsten data[157)] shown in Figs. 33 and 34. The sticking probability is very dependent upon the crystal plane which is exposed to the gas. The sticking probability for N_2 on W[100] is initially almost independent of coverage and equal to ~0.6. In contrast, the sticking probability is almost zero for W(110) and decreases linearly from an initial value of ~0.1 for W(111). Each plane behaves in a different manner. Moreover, the sticking probability is very temperature-dependent as is shown in Fig. 34. Furthermore, the behaviour of sticking probabilities is likely to be more complex in a discharge where vibrationally and rotationally excited molecules exist.[157 a)]

Table 3. Classification of metals and semi-metals based on adsorption properties. (A indicates adsorption, NA no adsorption.) (From Ref.[156)])

Group	Metals	Gases						
		O_2	C_2H_2	C_2H_4	CO	H_2	CO_2	N_2
A	Ca, Sr, Ba, Ti, Zr, Hf, V, Nb, Ta, Cr, Mo, W, Fe,[a] (Re)	A	A	A	A	A	A	A
B_1	Ni, (Co)	A	A	A	A	A	A	NA
B_2	Rh, Pd, Pt, (Ir)	A	A	A	A	A	NA	NA
C	Al, Mn, Cu, Au[b]	A	A	A	A	NA	NA	NA
D	K	A	A	NA	NA	NA	NA	NA
E	Mg, Ag,[a] Zn, Cd, In, Si, Ge, Sn, Pb, As, Sb, Bi	A	NA	NA	NA	NA	NA	NA
F	Se, Te	NA	NA	NA	NA	NA	NA	NA

[a] The adsorption of N_2 on Fe is activated, as is the adsorption of O_2 on Ag films sintered at 0°.
[b] Au does not adsorb O_2.
[c] Metal probably belongs to this group, but the behaviour of films is not known.

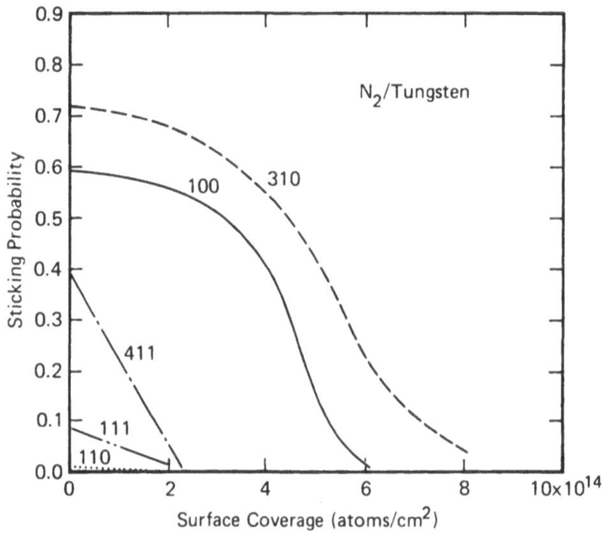

Fig. 33. Sticking probability for nitrogen on tungsten single crystal planes. T = 300 K. (From Ref.[157])

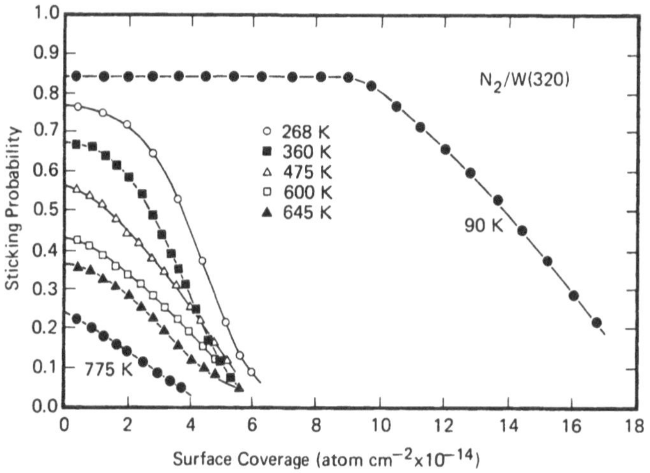

Fig. 34. Sticking probability for nitrogen on the [320] plane of tungsten. Temperature is a parameter. (From Ref.[157])

Despite the complexities, a knowledge of whether a particle reacts with a given surface can be very helpful in the interpretation of discharge phenomena. Two examples will now be presented. Consider an argon-nitrogen discharge in a system of the type shown in Fig. 1. Material is sputtered from the target and deposited upon the anode. With other parameters held constant, the amount of nitrogen incorporated into the deposited film depends upon the partial pressure of nitrogen in the discharge. However, the concentration of nitrogen in the film at a given partial pressure depends upon the type material which is being deposited (sputtered). Materials which have a high reaction probability for the chemisorption of molecular nitrogen have a high concentration of nitrogen in the deposited film[70]. Those materials which do not chemisorb molecular nitrogen have a small nitrogen concentration under similar discharge conditions[70]. The interpretation is quite clear. The predominant nitrogen species in the discharge are molecules, atoms, and ions with the concentration of N_2 being much larger than that of the other species. The mechanism for incorporation of nitrogen into the growing film is dominated by chemisorption of the molecule in situations where this is probable. This is a direct consequence of the fact that the incident flux of molecules is greater than the flux of the other species. When molecules do not react, the dominant mechaniam involves the chemisorption of ions or atoms. This of course requires a higher partial pressure of nitrogen in order to obtain an equivalent flux of atoms to the surface. (It should be noted that chemisorption followed by burial is probably the dominant means by which particles are incorporated into a growing film in a plasma situation.)

Another example involves plasma etching where gas phase species from a plasma interact with solid materials to produce volatile compounds[158]. An example would be the following reaction:

$$CF_4(\text{gas}) + \text{Plasma} + \text{Silicon} \rightarrow \ldots \rightarrow SiF_4(\text{gas}) + \ldots .$$

It has recently been shown that several of the gases used in plasma etching (CF_4, CF_3H, CF_3Cl, CF_2Cl_2, CCl_4) do not chemisorb on silicon, silicon dioxide, or silicon nitride[131]. The relatively inert nature of these halocarbon gases with respect to silicon and its compounds tends to elucidate the role of the glow discharge, i.e., *the discharge generates reactive radicals by dissociating the halocarbon molecule.* The radicals subsequently react with the surface to form volatile products. A knowledge that the ground state molecule does not react greatly facilitates interpretation in this instance.

4.2 The Interaction of Radicals and Atoms with Surfaces

Experiments and/or theory from the field of surface science offer practically no guidance to investigators working with plasmas about the interaction of radicals with surfaces. Sticking probabilities and recombination cocficients for radicals on well characterized surfaces are virtually unknown. To our knowledge, there is no instance where it has been determined that the chemisorption of a molecular radical is either dissociative or non-dissociative. In most instances, it is not even certain whether or not a given radical will be chemisorbed on the surface of interest. Moreover, sticking probabilities, recombination coefficients, etc. will depend strongly upon the composition and structure of the surface. The production of damage in the surface region or the adsorption of impurity gas is expected to substantially affect these parameters. Therefore, data obtained for uncharacterized surfaces cannot be used confidently to predict behavior in plasma situations. Our approach to this section is dictated by the lack of information described above. Therefore, we will only attempt to present a few somewhat intuitive guidelines for the interpretation of data obtained in a glow discharge environment.

Radicals frequently chemisorb on surfaces which appear inert to the parent molecule. This conclusion is not surprising but deserves some discussion. For example, nitrogen atoms are chemisorbed on nickel[159], copper[160,161], iridium[162] and rhodium[162] whereas the nitrogen molecule is not chemisorbed on any of these surfaces. Moreover, even for surfaces which do chemisorb molecular nitrogen, the surface concentration can often be increased by exposure to atoms[133,163]. In addition, new binding states on a given surface are often populated[133,163] by exposure to atoms. Other simple diatomic molecules (O_2, H_2, etc.) appear to behave in a manner similar to nitrogen. For example, F_2 reacts slowly with a silicon surface[164,165] (reaction probability $\sim 10^{-6}$ to 10^{-5}) whereas evidence from plasma etching experiments suggest that F atoms rapidly attack silicon[166]. Similar conclusions about the reactivity of radicals can be derived from experiments performed in our laboratory[167]. Relatively inert gases (N_2, CH_4, C_2H_6, CF_4, CF_3H, C_2F_6, C_3F_8) are introduced into closed volume of a well-baked, ultra-high vacuum system. The gases in their ground state do not react with the vacuum system walls and therefore the pressure in the closed volume remains relatively constant and no impurity gases are formed. An electron beam is then used to dissociate the molecules and the pressure decreases. Three fates for the radicals generated by the electron beam are possible: they can be adsorbed on the surrounding surfaces, they can remain in the gas phase as radicals, or they can recombine to form stable gases. None of the radicals remained in the gas phase and recombination was only important for N_2. Therefore, it is clear that radicals such as N, CH_3, CH_2, CF_3, CF_2, C_2F_5, C_2H_5, etc. are strongly chemisorbed on the surfaces of a typical vacuum system. These surfaces are usually quite inert. (The well-known phenomena of ion gauge pumping is partially related to the reactivity of these radicals.) These results lead to our first suggested guideline for the interpretation of data obtained in a plasma. *It should be assumed that a radical will react with a given surface unless there is strong evidence to the contrary.*

There are two mechanisms which allow a radical to react in situations where the parent molecule is inert. First, the reaction between a surface and parent molecule may be exothermic but require a large activation energy to proceed. (This is probably the case for methane which is very unreactive toward a large variety of clean surfaces[168].) When the activation energy is provided by dissociating the molecule while it is in the gas phase, then the resulting fragments spontaneously react with the surface.

The second mechanism is related to adsorbed states which are endothermic. Consider, for example, a diatomic molecule whose dissociation energy is D and where the binding energy of each atom to the surface is E_a. Dissociative chemisorption will only occur if $2E_a > D$. The dissociation energy of the molecule must be provided by the adsorption energy of the resulting atoms. For the case of N_2, E_a must be greater than ~ 4.8 eV. That is, nitrogen molecules will not dissociatively chemisorb on a surface where the binding energy for each atom to the surface is less than 4.8 eV. It is expected that there are many surfaces where the binding energy is less than this value. Therefore, nitrogen atoms could be strongly chemisorbed on this type of surface whereas nitrogen molecules would not react.

The sticking probability of radicals is often large but not unity. Nitrogen atoms have been shown to chemisorb on iridium and rhodium even though the source of the atoms was not visible to the sample[162]. The present author has also observed CF_3 and CF_2 radicals mass spectrometrically in a situation where the particles traveled from the plasma through a valve and subsequently through about 50 cm of 2.54 cm stainless-steel tubing. They must have made numerous wall collisions before detection. In addition, many authors have generated radicals in a plasma and then observed the reaction downstream. Furthermore, the fact that the presence of one surface decreases the reaction rate at another surface within the plasma (the so-called loading effect) is a direct consequence of the fact that the sticking probability of radicals is not unity. The loading effect is a widely occuring phenomena in plasma etching situations. These observations lead to our second suggested guideline for the interpretation of plasma data. *In the absence of evidence to the contrary, it should be assumed that radicals have reasonably large sticking coefficients but that they are not unity.*

The importance of atoms in initiating and sustaining chemical reactions has long been recognized and a recent investigation for the hydrogen-carbon system by Balooch and Olander[169] emphasizes this point. They found that molecular H_2 was very unreactive toward graphite because dissociative chemisorption did not occur. However, atomic hydrogen generated in a hot tungsten oven reacted quite rapidly to produce either CH_4 or C_2H_2 depending upon the surface temperature. Their reaction probabilities are shown in Fig. 35 as a function of inverse temperature, 1/T. There are three distinct regions where different reactions

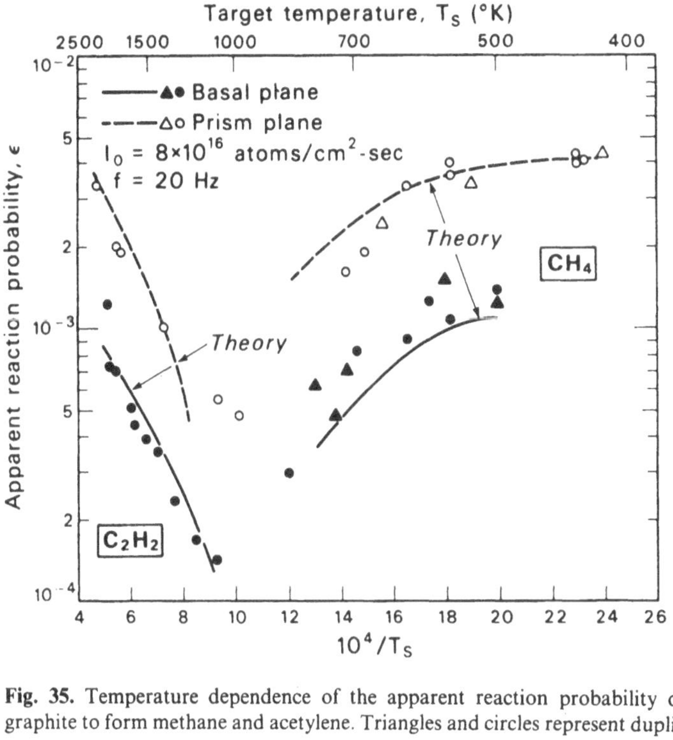

Fig. 35. Temperature dependence of the apparent reaction probability of hydrogen atoms with graphite to form methane and acetylene. Triangles and circles represent duplicate runs. (From ref.[169])

dominate i.e. methane is produced at low temperature, recombination with the formation of H_2 dominates at intermediate temperatures e.g. 1000 K, and acetylene is produced at high temperature. Atoms generated in a discharge would undoubtedly react in a manner similar to those generated by the hot oven in these experiments (see Gruen, et al., Vol. 0 of this series).

Recombination of radicals and ions to form stable gases is also important in discharges. For example, recombination reactions in a pure CF_4 discharge produce C_2F_6. When oxygen is added, recombination reactions also produce CO, CO_2 and COF_2. That is, some of the carbon which enters the system as CF_4 leaves the system in the form of volatile carbon-oxygen compounds. These recombination reactions may occur either in the gas phase or on surfaces and may dominate the chemistry which is occurring in the discharge. Unfortunately, we are aware of no information from the field of surface science which would make a significant contribution to the understanding of recombination reactions in a plasma environment.

4.3 The Interaction of Electronically Excited Molecules with Surfaces

In this section we deal with particles which are excited in the gas phase and then approach a surface. Excitation to optically allowed state are probably not important in this context since they will remain in the excited state for a period less than $\sim 10^{-7}$ sec. During this period they only travel ~ 1 mm. However, metastable molecules and ions have long lifetimes and therefore may be de-excited primarily at surfaces. The question, "Do these excited molecules react differently from ground state molecules?" is the topic of this section. Since little information is available, the discussion is highly speculative.

It is well known that electronic excitation of molecules can strongly affect chemical reactions which occur in the gas phase. One might therefore assume that similar processes would occur when electronically excited molecules approach a surface. However, we suspect that this is not the case. When an electronically excited molecule approaches a *metallic* or *semiconducting* surface, it is very efficiently de-excited by an Auger or resonant transition (see sect. 2.1). For slow particles, the de-excitation appears to occur before the particle begins to interact strongly with the surface. Therefore, from a chemical point of view, the incident particle is in its electronic ground state although it may well be vibrationally excited. Thus, a chemical reaction induced primarily by electronic excitation is not likely to occur or will occur only with small probability. The evidence for this conclusion is admittedly quite small and is essentially based upon the authors view as to what happens when an electronically excited molecule approaches a surface. However, there are two pieces of experimental evidence which tend to support this view. First, an N_2^+ ion, which has more than 15 eV of electronic excitation energy, does not react chemically with tungsten, molybdenum, or nickel surfaces unless it also has greater than 9 eV of kinetic energy (see Fig. 28). Nitrogen atoms with small kinetic energy will react with these same surfaces. The chemical reactivity of the N_2^+ ion is caused by its kinetic energy which leads to dissociation upon impact with the surface. The energy contained in electronic excitation is not by itself effective in causing a chemical reaction.

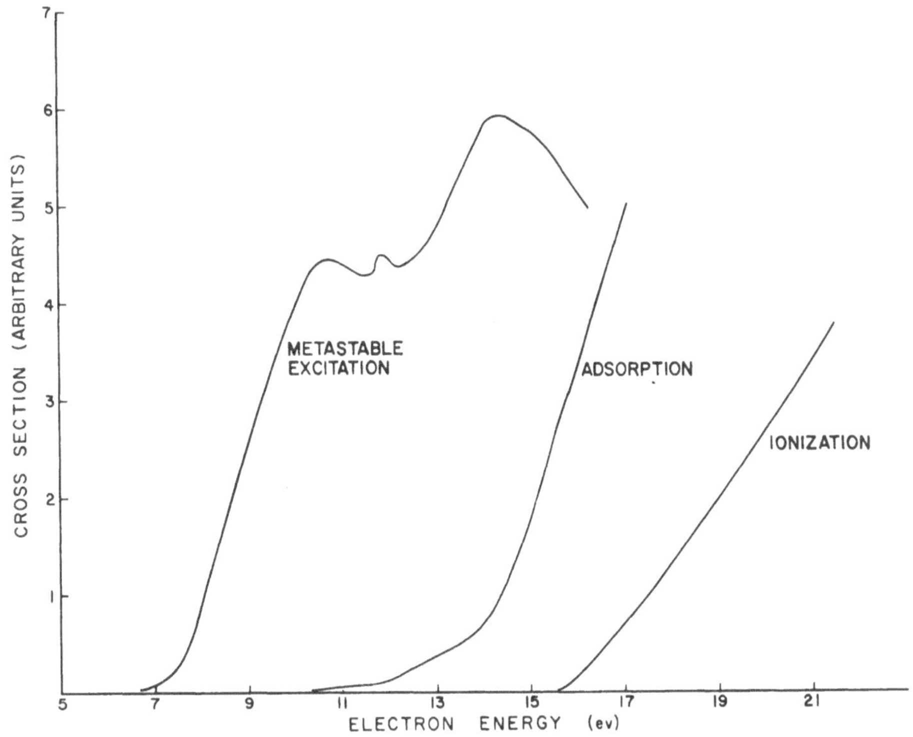

Fig. 36. Cross section versus Electron Energy. The cross section for excitation of metastable molecules is from ref.[170)] and for adsorption from ref.[171)]. The data was taken under very similar conditions in almost identical tubes

A similar conclusion can be drawn from the interaction of metastable nitrogen molecules (the $A^3\Sigma_u^+$ state) with these same surfaces. The relative excitation cross section (the excitation function) for this state is shown in Fig. 36 (see ref.[170)]). Direct excitation by electron impact has a threshold at approximately 6 eV and has a maximum at slightly higher energies. De-excitation from the $B^3\pi_g$, $C^3\pi_u$, etc. states also populate this level. The second curve in Fig. 36 is labeled "adsorption" and is the electron-induced-cross-section for a N_2 molecule to leave the gas phase i.e. react chemically at the surface. The metastable excitation cross section is large where the adsorption cross section is small e.g. at 10 eV. Moreover, there is no evidence from the shape of the adsorption cross section to indicate the presence of a contribution from metastables. It appears that metastable molecules are de-excited at the surface without causing a chemical reaction.

An opposite conclusion could be drawn for other types of material such as ionic solids or insulators. Strong evidence is beginning to appear in the literature[151)] which suggests a two-electron, *inter-atomic* Auger process can produce a positive ion at a position in the lattice where an negative ion originally existed. The positive ion now finds itself in a strongly repulsive Madelung potential and can be desorbed into the gas phase. This process should also presumably induce other types of chemical reactions. An *inter-atomic* process induced by ionization of a core level has also been observed to produce desorption[151 a)]. In these two sets of experiments, the core holes have been generated with electrons or photons. However, the present author fully expects that the de-excitation of ions or metastable neutrals at the surface of *some* ionic solids (and also at other types of surface) will be shown to produce chemical reactions and also desorption. The processes might be interpreted as sputtering, but would not be induced by momentum transfer from the incomming ion (neutral) to a surface atom.

Acknowledgments
The author gratefully acknowledges many helpful discussions with John Coburn.

5 References

1. Maissel, L.: Application of Sputtering to the Depositon of Films. In: Maissel, L., Glang, R. (eds.), Handbook of Thin Film Technology, New York, McGraw-Hill 1970 (review)
2. See for example, Vepřek, S.: Pure Appl. Chem. *48*, 163 (1976)
3. For a review see, Coburn, J. W., Winters, H. F.: J. Vac. Sci. Technol. *16*, 391 (1979)
4. Hudis, M.: Plasma Treatment of Solid Materials. In: Techniques and Applications of Plasma Chemistry, Hollahan, J. R., Bell, A. T. (eds.), New York, John Wiley & Sons 1974
5. Hagstrum, H. D.: Phys. Rev. *96*, 336 (1954)
6. Hagstrum, H. D.: Phys. Rev. *104*, 672 (1956)
7. Hagstrum, H. D.: Phys. Rev. *122*, 83 (1961)
8. Hagstrum, H. D., Takeishi, Y., Pretzer, D. D.: Phys. Rev. *139*, A526 (1965)
9. Probst, F. M., Lüscher, E.: Phys. Rev. *132*, 1037 (1963)
10. Probst, F. M.: Phys. Rev. *129*, 7 (1963)
11. Hagstrum, H. D.: J. Vac. Sci. Technol. *12*, 7 (1975)
12. Hagstrum, H. D.: Low Energy De-excitation and Neutralization Processes Near Surfaces. In: Inelastic Ion-Surface Collisions, Tolk, N. H., Tully, J. C., Heiland, W., White, C. W. (eds.), New York, Academic Press 1977
13. Schekhter, S. S.: J. Exptl. Theoret. Phys. (USSR) *7*, 650 (1937)
14. Hagstrum, H. D.: Phys. Rev. *123*, 758 (1961)
15. Hagstrum, H. D.: Phys. Rev. *104*, 317 (1956)
16. Medved, D. B., Mahadevan, P., Layton, J. K.: Phys. Rev. *129*, 2086 (1963)
17. Arifov, U. A., Rakhimov, R. R.: Izv. Akad. Nauk, SSSR Ser. Fyz. *24*, 657 (1960); English Translation in Bull. Acad. Sci. (USSR) Physics Ser. *24*, 666 (1961)
18. Dietz, L. A., Sheffield, J. C.: J. Appl. Phys. *46*, 4361 (1975)
19. Fano, U., Lichten, W.: Phys. Rev. Lett. *14*, 627 (1965)
20. Lichten, W.: Phys. Rev. *164*, 131 (1967)
21. Barat, M., Lichten, W.: Phys. Rev. *6*, 211 (1972)
22. Garcia, J. D., Fortner, R. J., Kavanagh, T. M.: Rev. Mod. Phys. *45*, 111 (1973)
23. Burhop, E. H. S., Asaad, W. N. In: Advances in Atomic and Molecular Physics, Bates, D. R. (ed.), Vol. 8, p. 163, New York, Academic Press 1972
24. Beuhler, R. J., Friedman, L.: J. Appl. Phys. *48*, 3928 (1977)
25. Schram, B. L., Boerboom, A. J. H., Kleine, W., Kistemaker, J.: Physica *32*, 749 (1966)
26. Sternglass, E. J.: Phys. Rev. *108*, 1 (1957)
27. Parilis, E. S., Kishinerskii, L. M.: Fiz. Tverd. Tela. *3*, 1219 (1960) [Sov. Phys. — Solid State *3*, 885 (1960)]
28. Ball, D. J.: J. Appl. Phys. *43*, 3047 (1972)
29. Chapman, B. N., Downer, D., Guimarães, L. J. M.: J. Appl. Phys. *45*, 2115 (1974)
30. See for example, Massey, H. S. W., Burhop, E. H. S. In: Electronic and Ionic Impact Phenomena, p. 149, London, Oxford University Press, 1952

31. Christophorou, L. G. In: Atomic and Molecular Radiation Physics, p. 35. London-New York, Wiley-Interscience, 1971
32. Rapp, D., Englander-Golden, P.: J. Chem. Phys. *43*, 1464 (1965)
33. See for example, Lindhard, J., Scharff, M., Shiøtt, H. E.: Mat. Fys. Medd. Dan. Vid. Selsk. *1963*, 33 Nr. 14
34. Firsov, O. B.: Zh. E.T.F. *36*, 1517 (1959), [Transl. Sov. Phys. JETP *9*, 1076 (1959)]
35. Sigmund, P.: Phys. Rev. *184*, 383 (1969)
36. Thompson, M. W.: Phil. Mag. *18*, 377 (1968)
37. Bøttiger, J. et al.: Rad. Effects *11*, 69 (1971)
38. Sigmund, P., Matthies, M. T., Phillips, D. L.: Rad. Effects *11*, 39 (1971)
39. Taglauer, E., Heiland, W.: Nucl. Instru. Methods, *132*, 535 (1976)
40. Poelsema, B., Verhey, L. K., Boers, A. L.: Surf. Sci. *55*, 445 (1976)
40a. Karpuzov, P. S., Yurasova, V. E.: Phys. Status Solidi *B47*, 41 (1971)
41. Robinson, M. T., Torrens, I. M.: Phys. Rev. *B9*, 5008 (1974)
42. Harrison, D. E., Moore, W. L., Holcombe, H. T.: Rad. Effects *17*, 167 (1973)
43. Harrison, D. E. et al.: J. Appl. Phys. *39*, 3742 (1968)
44. Bohr, N.: Mat. Fys. Medd. Dan. Vid. Selsk. *1948*, 18 Nr. 8
45. Lindhard, J., Nielsen, V., Scharff, M.: Mat. Medd. Dan. Vid. Selsk. *1968*, 36 Nr. 10
46. Lindhard, J. et al.: Mat. Fys. Medd. Dan. Vid. Selsk. *1963*, 33 Nr. 10
33 Nr. 10
47. Carter, G., Colligon, J. S.: Ion Bombardment of Solids, p. 8, New York, Elsevier, 1968
48. Hasted, J. B.: Physics of Atomic Collisions, p. 342, New York, Butterworths 1964
49. Schiff, L. I.: Quantum Mechanics, p. 281, New York, McGraw-Hill 1955
50. Firsov, O. B.: J. Exp. Theor. Phys. *33*, 696 (1957); [English Trans.: Sov. Phys. JETP *6*, 534 (1958)]
51. Gombas, P.: Handb. Physik *36*, 109 (1956)
52. Abrahamson, A. A.: Phys. Rev. *178*, 76 (1969)
53. Goldstein, H.: Classical Mechanics, p. 73, London, Addison-Wesley, 1959
54. Weissman, R., Sigmund, P.: Rad. Effects *19*, 7 (1973)
55. See for example, Ref. *33*, p. 19
56. Bøttiger, J., Winterbon, K. B.: Rad. Effects *20*, 65 (1973)
57. Oen, O. S., Robinson, M. T.: Nucl. Inst. Methods *132*, 647 (1976)
58. Eckstein, W., Matschke, F. E. P., Verbeck, H.: J. Nucl. Materials *63*, 199 (1976)
59. Smith, A. G., Carter, G.: Rad. Effects *12*, 63 (1972)
60. Kornelsen, E. V., Sinha, M. K.: J. Appl. Phys. *40*, 2888 (1969)
61. Kornelsen, E. V., Sinha, M. K.: J. Appl. Phys. *39*, 4546 (1968)
62. Kornelsen, E. V., Can. J. Phys. *42*, 364 (1964)
63. See Ref. 47, p. 363
64. Kay, E., Winters, H. F. In: Trans. 3rd Internat. Vacuum Cong. Vol. 2, pt. 2, p. 351, New York, Pergamon, 1965
65. Kornelsen, E. V. et al.: Phys. Rev. *136*, A849 (1964)
66. Winters, H. F., Kay, E.: J. Appl. Phys. *38*, 3928 (1967)
67. Davis, W. D., Vanderslice, T. A.: Phys. Rev. *131*, 219 (1963)
68. Heim, G., Kay, E.: J. Appl. Phys. *46*, 4006 (1975)
69. Hoffmeister, W., Zuegel, M.: Thin Solid Films *3*, 35 (1969)
70. Winters, H. F., Kay, E.: J. Appl. Phys. *43*, 794 (1972)
71. Schwartz, G. C., Jones, R. E.: IBM J. Res. Dev. *14*, 52 (1970)
72. Winters, H. F., Horne, D. E.: Phys. Rev. *10* B, 10 (1974)
73. Güntherschulze, A., Meyer, K.: Vacuum *3*, 360 (1953)
74. Wehner, G. K.: Adv. Electronics Electron Phys. *7*, 239 (1955)
75. Behrisch, R.: Ergebn. Exact. Naturwiss. *35*, 295 (1964)
76. Kaminsky, M.: Atomic and Ionic Impact Phenomena on Metal Surfaces, Berlin-Heidelberg, Springer-Verlag, 1964
77. MacDonald, R. J.: Adv. Phys. *19*, 457 (1970)
78. Tsong, I. S. T., Barber, D. J.: J. Matr. Sci. *8*, 123 (1973)
79. Oechsner, H.: Appl. Phys. *8*, 185 (1975)
80. Winters, H. F. In: Radiation Effects on Solid Surfaces, Kaminsky, M. (ed.), Washington D.C., Amer. Chem. Soc., 1976

81. McCracken, G. M.: Rep. Prog. Phys. *38*, 241 (1975)
82. Sigmund, P.: Rev. Roum. Phys. *17*, 823 (1972)
83. Ibid. 969
84. Ibid. 1079
85. Brandt, W., Laubert, R.: Nucl. Inst. Methods *47*, 201 (1967)
86. For sublimination energies see, Honig, R. E.: RCA Rev. *23*, 567 (1962)
87. Weissman, R., Behrisch, R.: Rad. Effects *19*, 69 (1973)
88. Oechsner, H., Schoof, H., Stumpe, E.: Surface Sci. *76*, 343 (1978)
89. Laegreid, N., Wehner, G. K.: J. Appl. Phys. *32*, 365 (1961)
90. Stuart, R. V., Wehner, G. K.: J. Appl. Phys. *33*, 2345 (1962)
91. Bay, H. L., Roth, J., Bohdansky, J.: J. Appl. Phys. *48*, 4722 (1977)
92. KenKnight, C. E., Wehner, G. K.: J. Appl. Phys. *35*, 322 (1964)
93. Rosenberg, D., Wehner, G. K.: J. Appl. Phys. *33*, 1842 (1962)
94. Furr, A. K., Finfgeld, C. R.: J. Appl. Phys. *41*, 1739 (1970)
95. Southern, A. L., Willis, W. R., Robinson, M. T.: J. Appl. Phys. *34*, 153 (1963)
96. Oechsner, H.: Z. Phys. *238*, 433 (1970)
97. Vossen, J. L., Cuomo, J. J. In: Thin Film Processes, Vossen, J. L., Kern, W. (eds.), New York, Academic Press 1978
98. Wehner, G. K., Hajicek, D. J.: J. Appl. Phys. *42*, 1145 (1971)
99. Hanak, J. J., Pellicane, J. P.: J. Vac. Sci. Technol. *13*, 406 (1976)
100. Cuomo, J. J. et al.: IBM J. Res. Develop. *21*, 580 (1977)
101. Cuomo, J. J. et al.: J. Vac. Sci. Technol. *15*, 281 (1978)
102. Davis, W. D., Vanderslice, T. A.: Phys. Rev. *131*, 219 (1963)
103. Coburn, J. W., Kay, E.: J. Appl. Phys. *43*, 4965 (1972)
104. Vossen, J. L.: J. Electrochem. Soc. *126*, 319 (1979)
105. Vossen, J. L.: J. Appl. Phys. *47*, 544 (1976)
106. Coburn, J. W.: J. Vac. Sci. Technol. *13*, 1037 (1976)
107. See for example, Ziegler, J. F., Cuomo, J. J., Roth, J.: Appl. Phys. Lett. *30*, 268 (1977)
108. Yonts, O. E., Harrison, D. E.: J. Appl. Phys. *31*, 1583 (1960)
109. Hasseltine, E. H., Hurlbut, F. C., Olson, N. T.: J. Appl. Phys. *38*, 4313 (1967)
110. Jones, R. E., Winters, H. F., Maissel, L. I.: J. Vac. Sci. Technol. *5*, 84 (1968)
111. Chuang, T. J., Brundle, C. R., Wandelt, K.: Thin Solid Films *53*, 19 (1978)
112. Kim, K. S. et al.: J. Elect. Spec. Relat. Phenom. *5*, 351 (1976)
113. Patterson, W. L., Shirn, G. A.: J. Vac. Sci. Technol. *4*, 343 (1967)
114. Winters, H. F., Raimondi, D. L., Horne, D. E.: J. Appl. Phys. *40*, 343 (1969)
115. Anderson, G. S.: J. Appl. Phys. *40*, 2884 (1969)
116. Gillam, E.: J. Phys. Chem. Solids *11*, 55 (1959)
117. Shimizu, H., Ono, M., Nakayama, K.: J. Appl. Phys. *46*, 460 (1975)
118. Tarng, M., Wehner, G. K.: J. Appl. Phys. *42*, 2449 (1971)
119. Seitz, F., Koehler, J. S. In: Solid State Physics, Vol. 2, p. 333, Seitz, F., Turnbull, D. (eds.), New York, Academic Press, 1956
120. Harrison, D. E., Johnson, J. P., Levy, N. S.: Appl. Phys. Lett. *8*, 33 (1966)
120a. Andersen, H. H.: Appl. Phys. *18*, 131 (1979)
121. See Ref. 33, p. 27
122. Hart, R. R., Dunlap, H. L., Marsh, O. J.: J. Appl. Phys. *46*, 1947 (1975)
123. Maissel, L. I., Schaible, P. M.: J. Appl. Phys. *36*, 237 (1965)
124. See Refs. 70 and 114 for a discussion of the older literature
125. Winters, H. F. and Coburn, J. W.: Appl. Phys. Lett. *34*, 70 (1979)
126. Hosokawa, N., Matsuzaki, R., Asamaki, T.: Japan J. Appl. Phys. Suppl. 2, Part 1, 435 (1974)
127. Holland, L., Ojha, S. M.: Vacuum *26*, 53 (1976)
128. Holland, L., Ojha, S. M.: Vacuum *26*, 233 (1976)
129. Coburn, J. W., Winters, H. F.: J. Appl. Phys. *50*, 3189 (1979)
130. Vepřek, S., Webb, A. P., Oswald, H. R.: J. Nucl. Matr. *68*, 32 (1977)
131. Winters, H. F.: J. Appl. Phys. *49*, 5165 (1978)
132. Mauer, J. L. et al.: J. Vac. Sci. Technol. *15*, 1734 (1978)
133. Winters, H. F., Horne, D. E.: Surface Sci. *24*, 587 (1971)

134. Teloy, E. In: Trans. Third Intern. Vacuum Congr., Vol. 2, pt. 3, p. 613, Adam, H. (ed.), New York, Pergamon 1967
135. Winters, H. F.: J. Appl. Phys. *43*, 4809 (1972)
136. See for example, Coburn, J. W., Winters, H. F., Chuang, T. J.: J. Appl. Phys. *48*, 3532 (1977)
137. Hickmott, T. W.: Appl. Phys. Lett. *15*, 232 (1969)
138. DiMaria, D. J., Weinberg, Z. A., Aitken, J. M.: J. Appl. Phys. *48*, 898 (1977)
139. Curtis, O. L., Srour, J. R., Chiu, K. Y.: J. Appl. Phys. *45*, 4506 (1974)
140. Sickafus, E. N.: Phys. Rev. *16B*, 1436 (1977)
141. Wolff, P. A.: Phys. Rev. *95*, 56 (1954)
142. Amelio, G. F.: J. Vac. Sci. Technol. *7*, 593 (1970)
143. Bennett, A. J., Roth, L. M.: Phys. Rev. *5B*, 4309 (1972)
144. Hackenberg, O., Bauer, W.: Adv. Electron. Electron Phys. *11*, 413 (1959)
145. Birkhoff, R. D.: Handb. Phys. *34*, 53 (1958)
146. Kollath, R.: Handb. Phys. *21*, 232 (1956)
147. Madey, T. E., Yates, J. T.: Surface Sci. *63*, 203 (1977)
148. Menzel, D.: Surface Sci. *47*, 370 (1975)
149. Drinkwine, M. J., Lichtman, D.: Progr. Surface Sci. *8*, 123 (1977)
150. For gas phase dissociation cross sections see:
    a. Christophorou, L. G., Stockdale, J. A. D.: J. Chem. Phys. *48*, 1956 (1968)
    b. Rapp, D., Englander-Golden, P., Briglia, D. D.: J. Chem. Phys. *42*, 4081 (1965)
    c. Winters, H. F.: J. Chem. Phys. *44*, 1472 (1966)
    d. Winters, H. F.: J. Chem. Phys. *63*, 3462 (1975)
    e. Winters, H. F.: Chem. Phys. *36*, 353 (1979)
151. Knotek, M. L., Feibelman, P. J.: Phys. Rev. Lett. *40*, 964 (1978)
151a. Franchy, R., Menzel, D.: Phys. Rev. Lett. *43*, 865 (1979)
152. Overeijnder, H., et al.: Rad. Effects *36*, 63 (1978)
153. Kirby, R. E., Lichtman, D.: Surface Sci. *41*, 447 (1974)
154. Thomas, S.: J. Appl. Phys. *45*, 161 (1974)
155. Carriere, B., Lang, B.: Surf. Sci. *64*, 209 (1977)
156. Bond, G. C.: Catalysis by Metals: New York: Academic Press, 1962, p. 66
157. Singh-Boparai, S. P., Bowker, M., King, D. A.: Surface Sci. *53*, 55 (1975)
157a. Capitelli, M., Molinari, E.: Topics Curr. Chem. *90*, 59–109. Berlin · Heidelberg · New York: Springer: 1980
158. For a review of this subject see Coburn, J. W., Winters, H. F.: J. Vac. Sci. Technol. *16*, 391 (1979)
159. Gregory, J. C., Hayward, D. O. In: The Structrure and Chemistry of Solid Surfaces, p. 56-1, Somorjai, G. A. (ed.), New York, John Wiley & Sons, 1969
160. Tibbetts, G. C.: J. Chem. Phys. *70*, 3600 (1979)
161. Lee, R. N., Farnsworth, H. E.: Surface Sci. *3*, 461 (1965)
162. Mimeault, V. J., Hansen, R. S.: J. Phys. Chem. *70*, 300 (1966)
163. Dawson, P. T., Peng, Y. K.: Surface Sci. *33*, 565 (1972)
164. Kuriakose, A. K., Margrave, J. L.: J. Phys. Chem. *68*, 2671 (1964)
165. Chen, M., Minkiewicz, V., Lee, K.: J. Electrochemical Soc. *126*, 1946 (1979)
166. Mogab, C. J., Adams, A. C., Flamm, D. L.: J. Appl. Phys. *49*, 3796 (1978)
167. See Refs. 150d and e
168. Winters, H. F.: J. Chem. Phys. *62*, 2454 (1975)
169. Balooch, M., Olander, D. R.: J. Chem. Phys. *63*, 4772 (1975)
170. Winters, H. F.: J. Chem. Phys. *43*, 926 (1965)
171. Winters, H. F., Horne D. E., Donaldson, E. E.: J. Chem. Phys. *41*, 2766 (1964)

FSC
www.fsc.org
MIX
Papier | Fördert
gute Waldnutzung
FSC® C083411

Kolibri 360 GmbH
Ferdinand-Jühlke-Straße 7
99095 Erfurt, Deutschland
produktsicherheit@kolibri360.de